Hot Science is a series explorin̶... ...and technology. With topics f... ...dark matter to gene editing, these are books for popular science readers who like to go that little bit deeper ...

AVAILABLE NOW AND COMING SOON:

Destination Mars:
The Story of Our Quest to Conquer
the Red Planet

Big Data:
How the Information Revolution
is Transforming Our Lives

Gravitational Waves:
How Einstein's Spacetime Ripples Reveal
the Secrets of the Universe

The Graphene Revolution:
The Weird Science of the Ultrathin

CERN and the Higgs Boson:
The Global Quest for the Building
Blocks of Reality

Cosmic Impact:
Understanding the Threat to Earth from
Asteroids and Comets

Artificial Intelligence:
Modern Magic or Dangerous Future?

Astrobiology:
The Search for Life Elsewhere in
the Universe

Dark Matter & Dark Energy:
The Hidden 95% of the Universe

Outbreaks & Epidemics:
Battling Infection From Measles to
Coronavirus

Rewilding:
The Radical New Science of Ecological
Recovery

Hacking the Code of Life:
How Gene Editing Will Rewrite Our Futures

Origins of the Universe:
The Cosmic Microwave Background
and the Search for Quantum Gravity

Behavioural Economics:
Psychology, Neuroscience,
and the Human Side of Economics

Hot Science series editor: Brian Clegg

BIOMIMETICS

How Lessons from Nature can Transform Technology

BRIAN CLEGG

ICON

Published in the UK and USA in 2023 by
Icon Books Ltd, Omnibus Business Centre,
39–41 North Road, London N7 9DP
email: info@iconbooks.com
www.iconbooks.com

ISBN: 978-178578-989-2
eBook: 978-178578-988-5

Typeset by SJmagic DESIGN SERVICES, India.

Printed and bound in the UK.

For Gillian, Chelsea and Rebecca

ABOUT THE AUTHOR

Brian Clegg's many books include *Dice World* and *A Brief History of Infinity*, both longlisted for the Royal Society Science Book Prize, and, most recently, *Ten Days in Physics That Shook the World*.

CONTENTS

CONTENTS

ACKNOWLEDGEMENTS

My thanks as always to the team at Icon Books, notably Duncan Heath. Thanks also to the companies and individuals who responded to my enquiries about their biomimetic products and concepts – you will find them detailed in the book.

BETTER BY NATURE 1

The way that humans interact with and change the world around us is so different in scale from most other organisms that it can be easy to forget that we are indeed animals that have evolved within nature, rather than existing separately as strange alien beings that have no connection with the world around us.

Of course, our ability to go beyond our natural limits using technology has transformed *Homo sapiens* from being just another great ape to the dominant species on the planet. We have managed to transcend the limitations of our preferred environment to make our lives longer, safer and more enjoyable than nature allowed. Technology has enabled us to do everything from gaining an understanding of the universe to travelling around the world and reducing the impact of a virus pandemic. At a more basic level, we can keep dry when it rains, warm when it's cold and provide food for far more humans than the environment would naturally sustain.

These days we are used to hearing about the downside of human existence. We berate ourselves for the negative impact we have had on the environment, whether through climate change or destroying wildlife habitats. And it's only right that we do take a greater concern for the stewardship of the Earth. However, to only see human activities through this negative filter is to miss out on the opportunities that science and technology offer to make life better.

Many of the ideas for new developments and technologies have their origin in human ingenuity – but if we were to discount what nature has on offer, we would be ignoring a vast source of possible new approaches and ideas. Human impact on the world is dramatic – yet what individual humans can do is still tiny in scale when compared with the forces and structures of nature. You only have to see the impact that a storm or an earthquake – or a virus – can have on our apparently well-planned lives to realise this.

In this book we will be investigating the way that nature can provide inspiration to our scientists and engineers – a process known as biomimetics. The term 'biomimetic' dates back to at least 1960 and literally means 'emulating biology' – 'mimetic' comes from ancient Latin and Greek terms meaning 'being able to imitate'. This was a more general term than the roughly contemporary 'bionic', which is largely limited to electronic and mechanical imitation of the natural world.

The adjective was expanded to become a noun, 'biomimetics', by 1970 and would come to dominate, not only because it sounded more impressive than the dated-feeling 'bionics' but in conscious avoidance of association with the 1973 TV show, *The Six Million Dollar Man* about the bionic astronaut Steve Austin. The word 'biomimetics' is now sometimes used in a wider sense than focusing purely on the capabilities of biology, taking in technological advances where the original inspiration came from many different aspects of nature, and that is how it will be used here.

In making use of biomimetics, we recognise that the Earth is a vast laboratory where the mechanisms of natural selection have enabled evolutionary solutions to be developed for a wide range of problems. The difficulties that nature has evolved to cope with are often not the exact same ones that we face – but we can piggyback on the immense scope of natural experimentation and redeploy a solution to our advantage.

If anyone should doubt the capabilities of evolutionary solutions, consider the humble housefly. These small insects, typically 5 to 7 millimetres (0.2 to 0.3 inches) long, are common wherever humans live around the world. Next time you see a housefly, try and catch it in your hand. The chances are that it will easily avoid your grasp. Our technology is amazing. Yet we are very far from being able to build a robot the size of a housefly that can fly, walk and evade an attempt to catch it. Nature has evolved solutions to problems that remain well beyond our technological grasp.

All too often, the natural and the artificial are presented as opposing concepts, where everything natural is wonderful and the artificial is a poor second best, or even positively harmful. We should remember that 'artificial' means 'made by human artifice' – by skill. It's not a bad thing. And raw nature can be a distinctly nasty place. Equally, though, it would be big-headed in the extreme to think that we are incapable of learning from the world around us. Not only is the laboratory of nature huge, it has been carrying out evolutionary experiments for billions of years. Most of these experiments end in failure – but there have been so many tried that enough have worked, and worked well, that we could spend many generations attempting to discover the transferable concepts derivable from nature.

To see how biomimetics can result in a simple yet highly practical solution to a problem, we will take a trip back in time to Switzerland in 1941. An electrical engineer named George de Mestral, at the time aged 34, was out hunting in Alpine woodland with his shotgun and his dog, Milka. As they headed for home, he noticed that his socks and coat, along with Milka's fur, had picked up seed heads of the burdock plant, known as burrs.

The ability of these burrs to hitch a ride was the result of one of evolution's many attempts to deal with a reproductive barrier faced by plants. Unlike animals, plants are static. By default, their seeds drop to the ground at the base of the plant. This is a problem because when the seeds germinate, they will be competing both with each other and the parent plant, attempting to access the same nutrients, sunlight and water. It would be far better for reproductive success if the young plants could put some space between themselves and their parent.

The way that evolution works is that traits that increase the chances that a species will successfully reproduce are more likely to be passed on to the next generation and so on. Plants that developed mutations that gave them a slight edge in dealing with this problem lived to pass these mutations on, making the traits increasingly dominant. This has happened in a multitude of ways to deal with the seed distribution problem. There are many plants, for example, that make use of the air to spread their seeds. This approach ranges from windblown seeds, such as the dandelion's delicate 'clock' seed head to the helicopter-like wings found on some tree seeds, such as the sycamore, which enable the seeds to fly many metres from the tree.

Other plants have made use of the mobility of animals to give their seeds a lift. Many do this by giving the seeds a sweet coating, making the package attractive to eat. As a result, the seeds get passed through the animal's digestive system to be deposited elsewhere. But there is an alternative approach. If a seed should happen to stick to an animal's coat, it can piggyback on the moving animal, dropping off at a later time when it has been carried away from its parent. And it is this approach that came to be used by the burdock.

The heads of the plant, carrying seeds, are covered in little spines. At first sight these pointy extrusions may seem to be a way of putting off animals from eating the plant – but on closer inspection, these are not straight, pointed defensive spikes. Each spine ends in a small, tightly curved hook. When an animal brushes past, these hooks catch on the animal's fur, pulling the seed head from the plant to later be deposited elsewhere.

For most people, these burrs might be interesting – or fun (I remember a game involving throwing them at people as a child, scoring points if the seed heads stuck) – but de Mestral saw a more interesting potential in them. What nature had developed was a way of getting two things to stick to each other without requiring stickiness. This meant that the adhesion would not deteriorate over time. Most sticky things lose their adhesive qualities if they are repeatedly stuck and removed. But unless the hooks get broken, a burr will continue to attach and re-attach itself to fibres indefinitely. At the same time, it's an attachment that isn't permanent. The seed heads are intended to come off the animal's fur in time.

To de Mestral, this combination of an ability to repeatedly attach without losing grip, yet being relatively easy to remove when required, suggested a new way to make a fastener. Two pieces of a material, one featuring burr-like hooks, the other suitable fibres, would attach firmly to each other, but could be easily re-opened by simply pulling them apart. Getting from that first idea to a finished product took a considerable time. De Mestral obtained a first patent in 1955 and Velcro was launched on the world at the end of the 1950s.

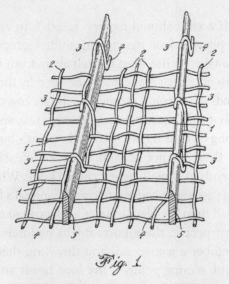

Fig. 1

From De Mestral's patent for Velcro.

George de Mestral (1958), *Separable Fastening Device*,
US3009235A, US Patent and Trademark Office

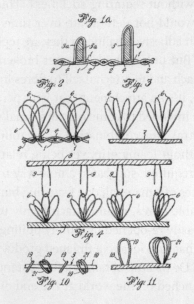

The original patent envisaged using two layers both lined with hooks: a 'separable fastening device', made up of 'two layers of woven fabric of the velvet type in which the loops have been cut to form hooks' – de Mestral had to invent a special device to do this cutting, based on barbers' clippers. However, by the time of his updated 1958 patent, de Mestral noted that to work effectively, 'the hooks of these layers of fabric are formed by a thread of artificial material, such as nylon', with their shape preserved by heat treatment. 'It has been found,' says the patent 'that the use of one layer of fabric of the hooked velvet type, as described above, with a layer of fabric of the loop type, such as terry or uncut velvet, provides greatly improved resistance to the separation of the two layers'.

De Mestral had improved on nature. The burrs attach to animal hairs. But a more secure attachment could be made if the hooks were paired with fabric covered in loops of fibre for the hooks to latch onto. The name de Mestral gave to his product was Velcro*, combining the French words for velvet (*velours*) and hook (*crochet*). This remains the registered trade name used by the company that developed the product, though other hook and loop fasteners now exist. Interestingly, in some languages, the invention is still directly linked to burdock in its name – in German, for example, it is

---

* The company descended from de Mestral's original organisation want their product to be known as 'VELCRO® Brand hook and loop fasteners' rather than Velcro, as the term tends to be generically attached to a wider range of hook and loop fasteners. They have attempted to make 'Don't say Velcro' go viral. When 'Velcro' is used in this text, please read 'VELCRO® Brand hook and loop fasteners'. Confusingly, though, the strangely plural company is called 'Velcro Companies' not 'VELCRO® Brand Companies'.

Klettverschluss (burdock fastener) and in Norwegian *borrelås* (burdock lock).

Sometimes, a product inspired by nature has a very limited application – often similar to the way it was initially used. But these hook and loop fasteners have found uses far beyond the initial idea of a way of fastening together two fabrics as an alternative to a zipper. The fasteners occur on everything from trainers to cable ties, and from splints to aircraft. They have also been used in space, whether for keeping tools in place inside spacecraft or for nose-scratching sticks fixed inside astronauts' helmets, leading to the frequently made, but entire false, assertion that Velcro was a spin-off benefit of the NASA space programme.

This, then, is biomimetics in action. A natural solution to a problem – how to be able to repeatedly fix two things together in a way that they can cleanly be separated – proved the inspiration for a product of human ingenuity. And the Velcro story fits perfectly with the narrative of biomimetics as a dramatic way of coming up with hugely successful new inventions that can help transform the world (even if it is in the less-than-Earth-shattering field of fasteners). Yet we will discover that the reality of biomimetics is often far more complex than this crude depiction – and what feel like transformative lessons from nature are often used once only, or never properly deployed at all. Understanding why this happens will be crucial to deciding whether biomimetics is a truly impressive concept, or an approach that is more show than substance.

Velcro was not the first invention to be inspired by nature, but before going deeper into the strengths and limitations of biomimetics, we need to take a step back and look at two key factors that come into play in the deployment of biomimetic solutions. The first is what evolution is and how it works – as this is primarily how the natural world develops solutions to problems in the first place. And the second is exactly what we mean by 'technology' and how it enables us to go beyond our evolutionary capabilities.

EVOLUTION AND TECHNOLOGY 2

Nature's problem-solving technique

For a long time, biology was something of a Cinderella science. It lacked a big idea – it was, in physicist Ernest Rutherford's cruel but accurate jibe about science other than physics, little more than 'stamp collecting'. Rather than explaining how and why biological organisms had become the way they were, biology simply catalogued and described.

The theory of evolution changed this. Like all the most dramatic aspects of science it is an apparently simple concept that explains a whole range of outcomes. (That 'apparently' is important. Doing science almost always involves admitting that things are more complex than we first thought.) In fact, the basics of evolution are so simple and inevitable that it's surprising it wasn't thought up far earlier.

If we only take the fundamental core of evolution, all we are saying is that if organisms vary within a species, and that variation can be passed on to offspring, then the variants that are more likely to survive are more likely to pass on their characteristics to future generations. That's it. Something that even the most fervent anti-evolution fanatic is likely to agree with.

Let's take a ludicrously simplified example. Imagine there's an environment in which heavy stones regularly rain down from the hills (it really doesn't matter why this is happening – it could be volcanic activity, aggressive birds or aliens having fun). And we have a species living there in which some individuals have got a strong shell, but some have a much weaker shell. Then it's pretty obvious that the organisms with the strong shells are more likely to survive long enough to have offspring. So, over time, more and more of the organisms will be of the hard-shelled variety.

Note that no intelligence is required in making this happen. There is no conscious design or intended direction of development. This was a random variation in the strength of the shells of the creatures.* And that, combined with passing on the characteristic of having a strong shell to the next generation, was enough to change the nature of the species to deal with the problem of falling stones.

This is how biological nature deals with difficulties. So, for example, in the previous chapter I mentioned that 'other plants have made use of the mobility of animals to give their seeds a lift. Many do this by giving the seeds a sweet coating, making the package attractive to eat' – it sounds as if the plants are consciously deciding to give their seeds a sweet coating. It is

---

* If (like me) you are fussy about the words you use, you may feel that 'creature' is a misleading word as it implies creation – which seems odd when the whole point here is that evolution is not about design. But it's useful to have an alternative word to organism, and that's all that I am implying.

very difficult to describe the process without it sounding like it is designed with a goal in mind. Nothing could be further from the truth as far as evolution is concerned. It involves repeated random variations, some beneficial, some harmful. Over the generations, the variants that are better able to survive pass on their abilities – and gradually solutions emerge.

This is often a slow, messy process. Although there are variants that can change details of an organism over a small timescale – think, for instance, of the frequent emergence of new variants of the SARS-CoV-2 virus that caused the Covid-19 pandemic – large changes, particularly in complex organisms, can take thousands or millions of years. But life has been around on Earth for several billion years, giving plenty of opportunity to develop solutions to challenging problems.

One particularly important point to understand is that evolution is not an optimising process. Because it is not aiming for a solution, but rather is simply wandering around randomly and happening upon an approach, nature's solutions are not necessarily the *best* way to do something. In fact, many biological structures and processes seem to work despite the way they are put together, based on what would be overly complex designs had they been designed. Nonetheless, given the vast scope of its laboratory in time, space and variation of environment, nature has managed to come up with many solutions to problems we could not hope to deal with using current technology.

The 'T' word

But what is 'technology'? It's what makes us distinctive as a species. Biologists tend to dismiss the concept of 'human exceptionalism' – the idea that humans are in some way special when compared with other organisms. Yet *Homo sapiens* has one huge distinction from other species in the way that we make use of technology to modify our abilities and the environment. This has arguably transformed the human being into a unique organism.

Some suggest this is still unwarranted exceptionalism, as other animals have been known to make use of tools. Chimpanzees, and a number of other apes, have been observed using rocks or twigs or leaves for a range of practical applications from missiles to a means of extracting tasty grubs from rotting logs. What's more, this kind of ability is not limited to primates – notably in the form of the New Caledonian crow. This canny bird is another user of twigs to retrieve otherwise inaccessible insects and has been observed intentionally producing a hooked end on a tool to make it more effective in rooting out a juicy bug.

However, highlighting the use of tools by animals other than humans is highly misleading. Firstly, the vast majority of primate species never use tools – and the few cases where they do have typically been one-offs, rather than a consistent ability that is found across the species. Even close relatives of chimps (and ours) such as gorillas and bonobos show very limited evidence of tool use. And secondly, this tool use does not in any way compare with the human use of technology because it is not transformative. Apes and crows use tools to be a little better at performing an existing task. Humans have used technology to give them abilities they never previously had, and to change their environment to make it more suitable for human life. As climate change demonstrates, we aren't necessarily very good at thinking through the long-term implications of those changes, but that doesn't undermine the transformative nature of our innovations.

When we use tools and technology, we span a whole spectrum of possibilities from slightly more sophisticated solutions to problems than the approaches chosen by apes and crows to taking on tasks that would be otherwise entirely impossible. Take a very simple example. Some while ago I was walking part of the Ridgeway, a national trail in the UK, on a blazing hot summer's day. Without the use of simple technology, this journey would have been sheer madness. It would just not have been safely possible. There was no water along the trail – the ground was bone dry. But I was able to keep going hour after hour because I carried water with me.

My water bottle achieved what nature took millions of years to develop in a camel – the ability to carry enough water to survive in an arid terrain. An unremarkable plastic bottle replaced a huge and complex biological transformation. This is already at least one step beyond, say, a chimpanzee using a leaf as a drinking vessel. The water bottle is something different in nature from the cupped hand that a leaf replaces because it could hold the water for as long as was required, while I walked, without taking up any of my attention.

Technology doesn't have to be technological

When technology is mentioned, we tend to think of complex devices, often based on electronics – but technology like the water bottle began to transform our abilities long before electrons were conquered as tools in the 20th century. Electronics inevitably brings to mind information and communication technology but, equally, such technology was transformative long before digital devices came on the scene.

Think, for a moment, of the impact of writing, a technology that dates back at least 5,000 years and was dreamed up independently in many early civilisations. The transformative power of writing lies in the ability to expand person-to-person communication beyond the here and now. Communication is a common trait in animals and even some plant species* – but it operates locally and ephemerally. My bookshelves, by contrast, contain written words from a book published today (I'm writing this on the day my book *Game Theory* was launched) through science fiction classics from the 1950s to Jane Austen's words from the early 19th century all the way back to works from over 2,000 years ago – whether we're talking Aristotle and Archimedes or the Bible.

Writing arguably benefited two main activities essential to the development of modern humans: storytelling and record keeping. Telling stories is far more than entertainment. We are, at heart, storytelling animals. Some even suggest it is the most distinctive of human capabilities. A book like this one is just as much a narrative as *Pride and Prejudice*. It is through story that we get an understanding of the world around us and of human nature. But the importance of narrative should not detract us from that record-keeping aspect.

In all probability, record keeping began with the need to identify ownership and account for transactions. And those are still activities that we undertake every day. However, the written word also enabled something far more important that

---

* There are a number of ways plants can communicate, usually as a response to stresses. Some, for example, tomato plants, release chemicals known as volatile organic compounds into the air, which are detected by other plants, while others, such as pea plants, interact in the soil via transfer of chemicals through their roots or, in the case of trees, via the networks of symbiotic fungi.

involved records – scientific discovery. Without the ability to record the results of experiments and scientific observation, our knowledge of the world would have remained limited to personal experience. And without the sharing of that information through writing, very little of the transformative technology that fills our world would ever have been developed, because this depends on building on the work of others.

Interestingly, although writing is not biomimetic – we weren't inspired by an existing natural mechanism for storing and transmitting information – we could think of it as bioconvergent if there is such a word. Nature does have its own information technology that dates back billions of years, putting our own timescales into proportion, in the so-called genetic code.

All living organisms, from humans down to bacteria and even the debatably alive viruses make use of DNA (or its cousin RNA) to pass on information from generation to generation. Those of us making use of the Roman alphabet have a code* of 26 different letters, plus a few formatting features such as punctuation, spaces and upper-case characters. In electronics, we usually use a binary format, where everything is represented by 0s and 1s. By comparison, in nature, the genetic code is quaternary, making use of bases: four chemical structures called adenine, cytosine, guanine and thymine, which are more conveniently referred to by their initial letters A, C, G and T.**

Sequences of these bases spell out the order in which to assemble building-block molecules called amino acids to form proteins, the essential chemicals of life. If ancient humans had known about the messages encoded in DNA, then this

---

* Strictly speaking all these things are ciphers, not codes. See my book *Conundrum*.
** These are the DNA bases – RNA replaces thymine with uracil (U).

could have been considered the biomimetic inspiration for writing – but they didn't, and it isn't. It is entirely possible for humans and wider nature to solve the same problem with similar types of solution, yet without one inspiring the other.

Even in nature, new developments are sometimes very similar yet not dependent on a common source of innovation. Take the eyes of vertebrates like us and cephalopods such as octopuses and squids. Each animal's eyes have considerable similarities. Yet all the evidence is that the vertebrates and cephalopods evolved their optical mechanisms separately. This could be down to either convergent or parallel evolution.

In convergent evolution, parts of two species evolve to have a similar appearance and/or function, but they aren't inherited from a common ancestor. A good example of convergent evolution is the wings of bats and birds. Parallel evolution is quite similar but happens when common ancestors of both species had a trait or traits that were then passed on to both species, even though the descendent species are otherwise different from each other. It is thought, for example, that some of the similarities between placental mammals and marsupials, from brain structures to fur, are the result of parallel evolution.

Without doubt, we are well placed to learn from nature and to be inspired by it. Though nature's information technology was not discovered until long after we had developed our own, we have, over the years, developed concepts in other areas of technology that have taken their inspiration from nature. Notably, this has been true in materials – substances to make things out of, food and medicine, structures and networks and optics.

Inventions and technology in each of these areas has been part of the human transformation of our environment and abilities. And arguably the first of the biomimetic technologies to be employed were biomimetic materials.

MATERIALS (STICKY OR OTHERWISE) 3

Materials science is arguably our most underrated scientific field, given how much impact it has had on everyday life. In English, the word 'material' is sometimes used to describe a fabric, but here we are dealing with a much wider use of the word as 'stuff out of which we can make things'. As chimps and crows have demonstrated, the simplest technology tends to involve simple tools. Similarly, many animals also build nests – sometimes very elaborate nests. In both cases, they are making use of natural materials – twigs or stones, for example. These materials are easy to find, but not necessarily ideal for dealing with the situations in which they are deployed. And so, among the first biomimetic developments by humans was the production of such tools from other materials, typically metals.

Biomimetic stone replacements

Metal use is likely to have begun with copper, which can be found isolated from minerals in nature – it is already part of the available natural resources and so cannot be considered biomimetic. However, the first practical biomimetic material to replace stone tools and weapons was bronze. This is an alloy – a mixture of two or more elements – in this case, formed from a mix of molten copper with either tin or arsenic. The artificial material bronze has a significant advantage over the naturally occurring copper because it is much harder, but it does not chip or crumble like the alternative natural material stone and can be formed into far more intricate shapes for specific tool (and weapon) use. This would commonly be as spear heads, axes and knives.

Arsenic was the first element to be alloyed with copper. The word 'arsenic' immediately conjures the idea of a natural poison, which makes it feel as if it should be a plant extract – but it is a metallic chemical element. Bronze dates back around 7,000 years. Tin was a later discovery, becoming more common in bronze than arsenic from around 4,000 years ago. Bronze would then be replaced in what would become an arms race of stone substitutions, with iron taking over a little over 2,000 years ago.

Similarly, branches, mud and rocks were used to build shelters, the precursor to our modern buildings, from the smallest hut to the largest cathedral. And it's true that wood and stone remain versatile parts of builders' toolkits. However, biomimetics would come into play in the development of artificial stones in the form of concrete and bricks.

Of these, bricks came first. In areas of clay soil, it was discovered that blocks of the clay dried in the sun made robust, flexible building blocks. Such bricks have been in use for at least 9,000 years. Around 3,000 years later, the first fired bricks – heated in a kiln to give them extra strength – were being produced, a form that would remain with subtle variations up to the present day. Bricks are very straightforward stone substitutes – but concrete would provide a whole step forward in imitating what stone could do in a far more flexible way.

Going liquid

Concrete is a significantly more modern invention than brick, though perhaps older than many would think with examples dating back over 3,000 years. By combining aggregate – typically small stones – with a mortar mix that sets hard, usually based on calcium oxide, concrete provided an equivalent to stone that could be made into any shape for which the builder could provide a mould.

By Roman times, concrete was already in use in monumental structures, such as the dome of the Pantheon in Rome. Later still, metal bars (typically steel) would be added as a kind of skeleton to form reinforced concrete, giving the extra strength required for modern concrete usage. Like bricks, concrete remains, in effect, a far more flexible mimic of natural stone. For this particular invention, though, there is a more dramatic modern twist that adds in a biomimetic lesson from biology.

Self-healing stone

One flaw that becomes apparent in many reinforced concrete structures over time is that they can develop cracks. Concrete is more susceptible than stone to shrinkage and to the impact of changes in temperature and the stresses caused by freezing and thawing. If concrete does crack, water can get through to the steel reinforcement within. As the metal rusts, it expands, contributing further to the damage. But what if concrete could learn to copy the way that nature deals with a similar issue?

We are all familiar with what happens when we cut a finger. Rather than retaining an open wound for the rest of our lives, the damage is repaired. We don't have to do anything to fix it (although, of course, it helps to keep the wound clean and infection-free) – our skin is self-healing. The same goes for broken bones, though intervention is often needed if they are to set cleanly. A number of possible biomimetic equivalents to self-healing have been devised, starting with the work of a microbiologist, Hendrik Jonkers of Delft University of Technology in the Netherlands, an establishment that specialises in cutting-edge technology.

In nature, many organisms produce solid structures from calcium carbonate – notably in the formation of eggs and shells, and of coral reefs. While it's not possible to directly employ the shell-building machinery of, say, a chicken or a clam, there are also bacteria that can generate calcium carbonate. Apart from their convenient size and easy deployment, the other benefit of using bacteria is that whatever the environment, from extremes of heat and cold to unpleasant chemical concentrations, it's possible to find a species that will thrive. This range of capabilities is important because the cement in concrete is sufficiently alkaline that many bacteria would struggle to survive. Jonkers was able to find an appropriate species, *Bacillus sphaericus*, which produces calcium carbonate if it has the right food, but that lives in the harsh surroundings of alkaline soda lakes.

Another capability of this versatile bacterium is to exist in spore form for long periods of time before 'coming to life'. Conventional spores of, say, a mushroom are a special form of offspring that can remain dormant for a long time. But some bacteria can reduce themselves to a spore-like form (technically known as endospores) as a survival mechanism when surroundings become hostile to life. Jonkers devised a bacterial spore and nutrient mix that could be included in concrete. Normally this stays dormant, but when water starts to get into the concrete, the spores turn back to fully functioning bacteria, which pump out calcium carbonate, repairing the cracks with this stone-like material.

As mentioned above, this technique is not the only approach in the development of self-healing concrete: a three-year project called Materials for Life was undertaken from 2015 by the Universities of Cardiff, Bath and Cambridge to compare the outcome of different self-healing techniques in a real-world setting on a live construction site on the A465 road in South Wales. The Bath team used the microbial approach, Cambridge employed so-called mineral healing agents delivered using microcapsules and Cardiff made use of both shape memory polymers and a technique known as a vascular flow network to deliver mineral healing agents.

The mineral healing agents used by both the Cambridge and Cardiff teams were based on sodium silicate. Such agents are existing products that are usually pushed into cracks in concrete to seal it, much as you might use a filler (spackle) in plaster. Here, though, to introduce the biomimetic self-healing aspect, there had to be some means to release the healing agents automatically from within the material. The Cambridge team used microcapsules, which are small polymer spheres that break when the concrete cracks. By contrast, the Cardiff flow networks are narrow channels that the healing agents are forced into under pressure, so that they will expand out into new cracks when the channels are breached.

Shape memory polymers, by contrast, are special plastics that can be moulded but return to their original shape when triggered by heat, light or water. Here, tendons of polyethylene terephthalate plastic were incorporated into concrete. When cracking occurs and water gets in, these tendons expand to fill the cracks. Meanwhile, for Bath's bacterial solution, *Bacillus pseudofirmus* was chosen, incorporated in spore form with yeast extract and 'calcium acetate' to act as food and raw materials.

In the trial, mock retaining panels of concrete incorporated different techniques (along with panels with no self-healing technology to act as controls). These panels were designed with a built-in weakness to encourage cracking. In the trial, all four technologies were successfully deployed and functioned, though different techniques seemed better with different types of cracking. At the time of writing, further research has yet to produce clear recommendations for the best approach to take for particular construction projects. The hope is that self-repairing concrete could become widely deployed in the future, but as is common with biomimetic technology, it is possible that this may remain a niche technique that isn't widely deployed in construction unless cost and reliability can be improved.

How long is a piece of string?

A similar development to the move from stone to the more versatile concrete has occurred with the development of ropes and metal cables. Originally, ropes were made from natural fibres, such as flax, hemp and papyrus, where the benefits of using multiple strands twisted around each other to enhance strength and reduce reliance on any single fibre were clear – the same approach would be taken in making ropes and cables from plastics and metals. The inspiration was certainly a naturally sourced rope, but the new alternatives were more durable and potentially much stronger. This development of biomimetic fibres has continued with the incorporation of new carbon-based fibres, giving strength comparable with steel at much lighter weight.

In modern times, we have seen biomimetic design take a different twist on fibre-based materials. For example, the 2018 World Motorsport Symposium awarded 'innovative product of the year' to a strong, thin material that made a double raid on nature for its design. This was a product for producing shaped panels based on fibre and resin. The predecessors of these panels were first fibreglass, where glass fibres are embedded in a resin matrix, and then carbon fibre panels where the glass is replaced by the much stronger carbon fibres.

In the 2018 innovation from Swiss manufacturer Bcomp, the fibres were themselves natural – flax. This plant has always been a versatile source. Its fibres are the basis for the fabric known as linen, while its seeds are the source of linseed oil. The fibre has also long been used in ropes, but here it was used to replace carbon fibres in resin panels. Flax may not be as strong as carbon fibre, but its production has significantly less environmental impact – and the motor-racing fraternity is eager to give their otherwise environmentally dubious sport a positive spin by making use of sustainable materials where possible.

More biomimetically significant (given that the fibres are natural), the flax is distributed in a grid given the brand name powerRibs, where the design is modelled on leaf veins. This, the manufacturers claim, gives the material greater stiffness to compete more effectively with carbon fibres. The actual design bears little resemblance to real leaf ribs – it's a net-like structure of flax on the underside of the panels. But the important aspect that is biomimetic is that, like those ribs that give leaves their strength, this is a three-dimensional structure that sticks out from the back of the thin layer of resin that forms the top surface. According to the manufacturer, this triples the stiffness, improves safety in crashes and results in up to 250 per cent higher vibration damping – while keeping weight to a minimum.

At the time of writing, this flax and resin material has limited applications. For example, Volvo combined it with recycled sea plastic to produce interior parts for a single demonstrator XC60 car. Similarly, it has been used for nine parts, including the roof panel, for the Porsche GT4 racing car. Even so, this material does seem to be one that has the potential to be wider used in manufacturing in the future.

Flax is a derived fibre – the original form in nature is in the bast or phloem – a layer of fibres that lie beneath the outer skin of the plant to give the stem its strength. However, another similarly inspiring material already occurs in nature in the form of fibres: the spider's web.

Silkily strong

Silk has been a valuable natural substance for thousands of years, produced from protein fibres that insects (notably silkworms) use to form their cocoons. However, traditional silk is arguably not the most interesting of naturally spun fibres. That accolade goes to spider silk, the material that is used to produce spiders' webs. Spider silk has three characteristics that are extremely valuable when combined – it is strong, light and stretchy. A thread of spider silk is half as strong as the equivalent thickness of mild steel, but it comes in at only around 1/8 the weight. And being several times more stretchy than the artificial fibre Kevlar, it has the potential to be ideal for the kind of applications where Kevlar is used, such as stab vests and flak jackets, where both strength and stretchiness are required to absorb energy from incoming weapons while minimising damage to the flesh beneath.

Although the silk industry has managed to make use of the natural fibres produced by silkworms, it's a messy and inefficient process – one that would be even harder to deal with if hoping to make use of natural spider silk. As a result, there has been a lot of interest in biomimetic fibres that are inspired by the spider's output. In essence, this means investigating both the structural composition of those spider fibres and the mechanisms that spiders use to spin their webs.

These nozzles, known as spinnerets, consist of large numbers of tiny extrusion devices that squeeze out the natural polymers and combine these ultra-thin strands to produce the desired type of silk depending on how the spider deploys its spinnerets: a single species can produce several different varieties of silk with, for example, different degrees of stickiness.

There have been a number of attempts to produce artificial fibres inspired by spider silk, a particularly strong home for developments being the University of Oxford. In 2001, Oxford researchers David Knight and Fritz Vollrath started a spin-off (appropriately enough) that went through a number of aliases as Spinox, then Deneflex, then Spinox again before becoming Oxford Biomaterials Limited. The original idea was to prototype spinning technology that could mimic spider spinnerets, but the company has moved on to look at developing spider silk-like fibres for medical device usage with fibres known as Spidrex.

This, however, is not Oxford's only spider silk spin-off. In 2020, a start-up called Spintex (names in this field are not highly imaginative) started work on developing a new silk-like protein as the feed for their own design of silk spinning devices. Fritz Vollrath crops up here again, along with Alex Greenhalgh and Martin Frydrych. Although the spinning mechanism is inspired by spiders, the aim here is to produce a thread that is an environmentally friendly artificial silk, rather than a fibre with the same extreme properties that are embodied in spider silk.

In the US and Canada, there has been considerable interest in the potential to replace Kevlar in flak jackets with the significantly stretchier spider silk. This was attempted not by entirely artificial means, but instead by making use of genetic modification to turn biological production of other proteins into the raw material of spider silk. In 2002, the Canadian company Nexia Biotechnologies claimed to have produced spider silk raw materials in the milk of genetically modified goats. Unfortunately, the end product did not live up to expectations and was dropped a few years later.

As yet, the spider's biomimetic inspirations have not become mainstream, but there remains considerable interest.

Best foot forward

Perhaps the second most famous use of a biomimetic material after the Velcro we met in Chapter 1 is derived from the remarkable ability of a type of lizard to walk up smooth walls and even to scale glass windows. Yet the story of the gecko's foot, and the gecko tape that was derived from it, has been far more typical of the general realities of biomimetics than the straightforward (if slow to develop) success of hook and loop fasteners.

Watching a gecko make its way up a smooth surface can be quite unnerving. It's as if Spider-man has got himself a lizard pet. Unlike a squirrel running effortlessly up a tree, making use of the interaction between claws and bark, the gecko has to have some way to attach itself to a smooth substance that offers no opportunity to either dig in or make use of an uneven surface for grip. Amazement with these animals' ability goes all the way back to the ancient Greek philosopher Aristotle, who remarked that geckos can 'run up and down a tree in any way, even with the head downwards'. But trees are only the starting point – as we have seen, geckos can walk up walls, across ceilings and down window glass.

Typically, we would expect the ability for something to attach itself to a flat, featureless surface to be dependent on either stickiness or suction cups, which make use of the pressure of the surrounding air to stay in place. And these were both considered in uncovering the secret of the gecko's foot. But this animal depends on a far more subtle mechanism – electromagnetism.

It's not that the gecko has magnets in its toes. Apart from anything else, this would only work on a suitably magnetic surface like iron – certainly not on glass or a plastered wall. But electromagnetism goes far beyond simple magnetic attraction. It is electromagnetism that prevents you from falling through a chair when you sit on it. Atoms are mostly empty space. What holds you up is not one set of atoms being directly supported by another set, but rather the electromagnetic repulsion between the atoms in the chair and those in your body.

That's a repulsive force as a result of particles with the same electrical charge being near each other. But electromagnetism is also responsible for the atoms in your body holding together. The bonds between atoms depend on electromagnetic attraction – the attraction that a positive and a negative electrical charge feel for each other. Simple electrostatic attraction – the mechanism by which, for instance, an electrically charged balloon can pick up bits of paper – does not explain a gecko's abilities, though. Its feet (and the walls it walks up) do not need to be electrically charged.

The van der Waals contribution

There is, however, a much weaker secondary electromagnetic attraction between atoms, known as the van der Waals force, which results from quantum effects when the electrons orbiting an atom briefly shift in position. This is a tiny force that we don't usually notice, yet it's the van der Waals force that appears to enable the gecko to perform its remarkable trick. And it can do this because its feet – and specifically its toes – are very different from those of any other lizard (or us, for that matter).

Take a look at a gecko's toes and you'll see that the bottom surface is covered in a series of horizontal pads called setae. Seen under a microscope, the setae look like collections of hairs, but in reality, they are covered in the confusingly named 'processes' – very thin extensions of the tissue of toe, which branch out into vast numbers of nanometre-scale bristles. These masses of tiny projections add up to a huge surface area that is in contact with the wall or other surface the gecko decides to encounter. And that's the secret of their glue-free adhesion. Because the gecko's setae are ideally structured to make the most of the van der Waals force.

Because of the strange quantum motion of electrons around the outside of an atom, the charge at any point undergoes small fluctuations – the van der Waals forces arise when these fluctuations pair up with opposite fluctuations in a nearby atom. And those atoms need to be very close indeed – the force is only detectable when two surfaces are less than

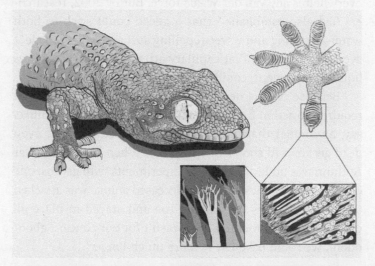

The Gecko's foot has multiple layers of fine structures, increasing the van der Waals attraction.

Matteo Gabaglio, CC BY-SA 3.0, via Wikimedia Commons

2 nanometres (billionths of a metre) apart. To give that some context, a typical human hair is around 75,000 nanometres thick. The result is a tiny attraction between each of the nanoscale protrusions on the foot and the surface as they come into close contact, attractions that add up over the whole of the foot to provide enough force to keep the gecko firmly in place.

There have been other theories to explain the ability of the gecko's foot, including a form of capillary attraction, but the van der Waals force has triumphed. Its role was reinforced because of one surface that geckos have a problem sticking to: the non-stick substance Teflon. This material, based on the synthetic polymer polytetrafluoroethylene (PTFE) has very low van der Waals attractiveness.* This is because the structure of PTFE has an outer coating of fluorine atoms, which tend to have a strong negative electrical charge that overwhelms any van der Waals force. But by 2002, researchers had also established that a gecko could stick to both water-attracting and water-repelling surfaces, which made it even more unlikely that capillary attraction, which requires liquid involvement, could be the mechanism.

Because of this dependence on simple attraction that requires no action from the gecko to make it happen (unlike, say, an animal that is clinging onto a tree branch), even dead geckos will stay firmly in place. When biologist Kellar Autumn was undertaking early experiments with geckos, one of his team's charges died. The deceased animal was attached to a sheet of glass by a single toe and stayed in place all day. Autumn commented 'if it wasn't for our concern about odour, we could have left it there much longer'.

---

* Oddly, due to a phenomenon that is not entirely understood, geckos *are* able to stick to Teflon if it's wet.

Remarkably, if every single protrusion on a typical gecko's foot was simultaneously in contact with a surface, that single foot could easily keep a heavy human in place – holding up to around 133 kg (290 pounds). In fact, the biggest problem a gecko has is not so much staying attached to a surface but getting its foot to peel off (more on this in a moment).

From feet to tape

Not surprisingly, there is a lot of interest in making use of the gecko's abilities in a biomimetic technology. After all, master this approach and you have a form of adhesion that is extremely powerful, yet doesn't deteriorate with repeated attaching and detaching as a conventional adhesive does. It seems to be an obvious biomimetic application. It certainly appealed to one of the brains behind the development of the wonder material graphene.

Physicist Andre Geim (co-winner with Konstantin Novoselov of the Nobel Prize for discovery of graphene) has something of a history of undertaking odd but interesting research. He first came to public attention in 2000 when he won the Ig Nobel Prize, a humorous award for research that 'first makes people laugh, then makes them think'. This win was for discovering that it was possible to levitate frogs using a strong magnetic field. (As Geim admitted, frogs were chosen because if they had ended up exploding, there would have been less fuss than if they had experimented on puppies.)

Later, with Novoselov, Geim would undertake 'Friday night experiments', when his team would be encouraged to try out something truly novel. As Geim put it, he thought there should be 'search, not re-search'. This was an opportunity to try out something new that would probably not work but was worth exploring – and might be fun. In 2003, such Friday night work led to a first possibility of producing a 'gecko tape', which Geim wrote up in the journal *Nature Materials* with details of a tape that could mimic gecko hair adhesion. This tape was described as self-cleaning and re-attachable.*

Reports at the time resulted in a degree of hyperbole, with *New Scientist*, for example, announcing that 'Gecko tape will stick you to the ceiling'. However, Geim – not usually one to undersell his work – was already warning that the product would not be easy to mass produce, and it wasn't clear how to make the artificial setae on the tape strong enough to withstand repeated application and peeling off. What's more, where a gecko's stickiness remains indefinitely, Geim's tape used flexible polymers in place of the stiff keratin that makes up gecko setae – which meant that the setae substitutes were too readily able to stick to each other, quickly rendering the tape unusable.

Four years later, researchers at the University of Akron in Ohio were describing a gecko technology sticky tape with four times the sticking power of a gecko's foot, meaning fully deployed gecko-sized pads could hold up around half a tonne. With pads like these on their feet, people would have no problem walking up walls – the only difficulty would be

---

* Interestingly, Geim was not sure about the rejection of capillary attraction as a mechanism for gecko adhesion and suggested at the time that the effect might involve both van der Waals and capillary forces.

managing to apply enough force to detach their hands and feet as they did so.

When a gecko walks up or down a surface, it moves its feet in two different ways. To attach itself, it pushes its foot straight into the surface. It can then hang from one foot because there is very strong adhesion pulling straight away from the surface. However, if the animal pulls its foot away at an angle, the setae detach easily. The gecko achieves this by flexing its toes as it peels the contact away from the toe-tips backwards, making it non-trivial to imitate a gecko in action. However, this is not a problem for a gecko-based tape or sticking plaster (band-aid) as the natural removal method for these is flexed peeling – for which there is practically no resistance.

In a typical gecko tape, the animal's setae are replaced by nanotubes of carbon fibre, which are attached to a sheet of flexible polymer, acting as the tape. The great thing about carbon nanotubes, which are effectively long, thin, flexible carbon crystals, is that they can be significantly narrower than the smallest protrusions from a gecko's foot. A typical nanotube has a diameter of a single nanometre, maximising the opportunity for van der Waals attraction. Within a year of the Akron announcement, researchers at the University of Dayton, Ohio (Ohio appears to be a particularly sticky state) were announcing a material with ten times the sticking power of the gecko's foot.

Such adhesive tapes are available commercially on a relatively small scale, offering the ability to stick under extreme temperature conditions and to surfaces that are wet or flexible that would defeat practically any conventional adhesive. You can now buy gecko tape on Amazon, say – but it has not become a widespread replacement for traditional adhesive tape. And that is for good reasons.

Although there have been some specialist developments trying to use variants of a gecko tape as grippers for robots, the general-purpose gecko tape falls down on several fronts. Geckos keep their feet clean, which may be due to secretions that sometimes leave behind gecko footprints on surfaces. Gecko tape tends to accumulate dust and grime on its fibres, rapidly rendering it ineffective because the pseudo-setae can't get close enough to the surface they are supposed to stick to. And, on a more mundane note, gecko tape is a lot more expensive than traditional adhesive tape, while there seems to be limited demand for a reusable product of this kind. Although it has niche uses, it is otherwise a solution looking for a problem.

Who wants to clean windows?

If geckos manage to produce stickiness from an unexpected source, another and arguably more beneficial natural example of evolution's random design school is a mechanism that is able to *prevent* stuff – specifically dirt – from sticking to a surface. While it's not entirely obvious that reusable sticky tape was something anyone was crying out for, it's certainly true that few people enjoy cleaning windows* – and in some cases, such as a tall, glass-faced buildings, it can be nearly impossible to get access to a piece of glass to clean it, requiring complex and expensive mechanisms to make this possible.

On the whole, windows are naturally 'washed' by rainwater – yet this doesn't mean that ordinary sheets of window

---

* With the exception of George Formby.

glass are self-cleaning. They still get dirty – because the dirt particles are sufficiently well attached to the surface of the glass that it needs more than exposure to rainfall to shift those fragments of filth. And the inspiration for a solution to this problem came from a lotus leaf.

Like a number of biomimetic discoveries, this was pretty much an accidental one, made by the German biologist Wilhelm Barthlott, who was an early user of the scanning electron microscope. These devices send a stream of electrons towards the surface of an object and detect the secondary electrons that are produced by the interaction of the initial stream with the atoms of the surface.

**A scanning electron microscope image
of pollen, showing the fine detail.**
Dartmouth College Electron Microscope Facility,
Public domain, via Wikimedia Commons

This produces a detailed picture of the surface of the object at high resolution.

In 1977, Barthlott became interested in the wide range of surprisingly complex structures that he found on the surfaces of both the leaves and seeds of plants. These look almost as if they are human-designed patterns when seen under the scanning electron microscope. To get a clear view of these microstructures, the plants needed to be carefully cleaned – but Barthlott noticed that some of the plants he studied never required any cleaning. This was the first hint of what would be termed (and even registered as a trademark as) the 'lotus effect'.

Generally speaking, when a drop of water lands on a surface, it spreads out. The object's surface becomes wet. But on a lotus leaf, each water drop forms a near-perfect sphere. This is because the attraction between water molecules is stronger than the attraction they feel to the leaf's surface. Water molecules are attracted to each other by hydrogen bonds, in some ways a similar type of effect to the van der Waals forces that give the gecko its grip. But hydrogen bonds are permanent, rather than transient, and provide a much stronger, longer-range attraction.

If it weren't for the hydrogen bonding in water, it would boil at less than –70 °C (–94 °F). This would mean no liquid water on Earth and no life. Thankfully, hydrogen bonding – which is caused by the attraction between the relatively positively charged hydrogen atoms and relatively negatively charged oxygen atoms in water – makes it much harder for water molecules to get away from each other than would otherwise be the case. Because the lotus leaf surface is covered with a large number of microscopic bobbles of a hydrophobic (water-repelling) waxy substance, the water molecules are more attracted to each other than they are to the surface. This prevents the water from spreading out and wetting the leaf.

The resultant near-spherical drops roll easily across the leaf surface. Usually, the tiny particles of dirt on a surface, such as a sheet of glass, are attracted to that surface more than they are to the water in a raindrop. But those microscopic leaf surface bobbles weaken the attraction to the dirt particles because there is less of the particle in contact with the surface. The rolling raindrop picks up the dirt particles in the same way that a rolled snowball picks up more snow. The dirt is accumulated by the raindrop and rolled off the surface, leaving the leaf clean.

After experimenting with a range of plants, Barthlott found that the best of the self-cleaners was the Indian or sacred lotus, *Nelumbo nucifera*. After a number of years trying to interest companies in the mechanism, in 1992, Barthlott registered the name 'Lotus-Effect' and applied for a patent in 1994. The patent was granted (in Europe) in 1998, and a paint product based on it called Lotusan, which had been in development in parallel with the patent application, was launched a year later. When the walls of a house are painted with Lotusan, they are less likely to become grubby over the years, provided they are exposed to rainfall.

This building paint has had moderate success, but arguably the bigger development was a method inspired by this mechanism that could be applied to the window cleaning problem. The leading glass British manufacturer Pilkington developed a product known as Pilkington Activ that is also self-cleaning. However, rather than directly mimicking the lotus leaf by preventing the surface from being wettable, this technology cleverly inverts the solution, *increasing* the wettability of the surface so that there is more interaction between rainwater and any debris on the glass.

The wetting effect is combined with a photocatalytic process that makes use of the compound titanium dioxide, more familiar as the UV-protecting particles in sunscreen. It had been known for some time that titanium dioxide enables light to help break down organic molecules. When such molecules land on windows, they make it harder to wash off grime. The titanium dioxide on the surface coating of the glass becomes electrically charged when hit by sunlight, an electrical charge that helps break down those sticky organic molecules by oxidising them, releasing the attached dirt.

Although the approach taken with Activ glass is not the same as the lotus effect, it can still be considered biomimetic because the self-cleaning inspiration came from the lotus, even if the mechanism didn't. Both Lotusan and Activ – and other products based on these approaches – have been reasonably successful. But as is often the case with biomimetics, because of their relative cost they remain relatively niche. Most householders looking to paint their house or install new windows are unlikely to ever come across these products. They have not matched the ubiquity of Velcro.

From geckos to frogs

Geckos are not the only animal life to give inspiration to material design – the latest poster animal to hit the headlines in 2017 was a tree frog. Like the gecko, this was a case of using a biomimetic approach to improve grip. But here there was a more significant application than glue-free sticky tape – one where grip can be a matter of life and death: in the design of tyre treads.

Tyre treads – the strangely designed grooves in the part of a vehicle tyre that comes in contact with the road – can seem something of an oddity. After all, they reduce the amount of rubber that is in contact with the road. This feels intuitively as if it should reduce the amount of grip that a tyre has, and in some circumstances it does. It seems strange that anyone would want a tyre that appears to be designed to be less safe.

If the road were perfectly dry, not having treads on a tyre makes perfect sense. That's why racing cars use 'slick' tyres without any treads on dry days. But the real world is not anywhere near as conveniently consistent as we would like it to be. We get rain. With a layer of water between rubber and the road, grip pretty much disappears. The tread enables a conventional tyre to push water out from under it, giving much better grip in wet conditions. (Though not on ice, which won't conveniently squeegee away.)

There is a very familiar natural analogue to tyre treads – your own fingers and toes. When you get in the bath, your finger and toe tips swell up as wrinkly pads. Many people think this is due to water being absorbed – but your fingers and toes are just as waterproof as the rest of your body.

Instead, what is happening is a response triggered by the nervous system. This is something your body does to itself (and that fails to occur with some types of nerve damage). It's always a little difficult to be sure with evolution what has caused a particular development, but it seems likely that the fingers and toes go this way in wet conditions to improve grip. Our hands and feet develop treads.

There is still some scientific dispute over this explanation. There have been experiments where people pick up various objects in circumstances where the object (a glass marble, for example) and the fingers can be wet and dry. In some cases, this seems to support the better grip in the wet argument, but the outcome is not conclusive. Even so, it is highly likely that grip comes into the survival of this trait.

Given the mechanism of developing wrinkles from the bath is a fairly recent discovery, it's unlikely that tyre treads were biomimetic from our fingers and toes. And, to date, most tread design has been notably engineering- rather than biology-based. But tree frogs have now started to make their mark. Arnob Banik, an engineering graduate teaching assistant at the University of Akron in Ohio (the state's reputation for stickiness continues), used the patterns on a tree frog's toe pads – its equivalent of our wrinkly bath toes – as inspiration for designing a 3D-printed rubber structure that produces improved grip in the wet.

Seen in close-up, the toe pads of some frogs have a scale-like pattern, where each scale is hollowed out a little like a suction cup. The shape seems to encourage wet adhesion – a combined effect of surface tension and the viscosity of trapped water. This is particularly the case on both tree frogs and a species known as a torrent frog that spends much of its time on wet stones near waterfalls. One tyre company, Continental, has produced tyres (ContiWinterContact) where each of the sections of the tyre in direct contact with the road (the bits that stand proud) has a surface with a shallow design based on the toe pads that give extra grip, though it would be imagined that the relatively shallow pads would not last long before being worn away.

Unlike many biomimetic products, these Continental tyres have become mainstream, as this design is easier to do here than with other products because there is no great increase in production costs. Having said that, it is notable that Continental does not make use of the biomimetic origin of the design in their sales material. This may be because comparative reviews do not seem to give the Continental tyres any significant benefit over other brands' winter tyres that do not use biomimetics.

The materials we have explored in this chapter play a special part in our lives, but arguably a less high-profile one than our next biomimetic concept: learning from nature to make better medicines and food.

PHARMACOPEIA AND FOOD

4

We are all used to foodstuffs being the product of nature, but in more recent times we have seen attempts to replace natural food with artificial equivalents, something that isn't often labelled biomimetic, but is probably the biggest intrusion of biomimetic design in everyday life. I certainly make use of products that contain, for example, artificial sweeteners most days, and I have been known to eat vegetable-based burgers that provide an increasingly convincing mimicry of meat.

As far as the pharmacopeia goes, until the 19th century, medical matters were not so much a science as a mix of guesswork, unsubstantiated theory and trial and error successes. It's arguable that most people treated by a doctor in, say, the 17th century suffered more harm than good. But one area of early medicine that did provide some positive results was the use of plant extracts to treat ailments.

Even though the basis of the early use of herbal preparations was often misguided (and sometimes still is today), there were clear successes. What was often the case was that a plant's natural defences were being repurposed to human advantage. Initially, this was a matter of directly using the plant itself, but true biomimetic development came into play when lessons were learned from the way that plant extracts worked to produce a similar, but improved effect. There are no better examples of this than in the development of aspirin.

Barking up the right tree

A very common medical issue is the need to suppress pain, reduce fever and limit inflammation. Particular plants and barks appear to have been used in this way stretching as far back as prehistory. We know that the bark of the willow tree was already listed as a necessary part of medical supplies during the Sumerian Third Dynasty of Ur, which takes us back about 4,000 years. Willow bark extract was also in use in ancient Egypt as a treatment for fevers, while the ancient Greek medical philosopher Hippocrates (he of the oath) singled out powdered willow bark as a go-to painkiller. Extracts from the plant spiraea, one of the plant species known as meadowsweet, were also used alongside the bitter willow bark to deal with fever and pain.

These pain-relief plants never totally dropped out of use, though their popularity waxed and waned. By the 18th century, though, no medic who achieved positive results would be without willow bark extract or meadowsweet syrup in his little black bag. As with most medications, popularity made the product more accessible, and it was during the 18th century that willow bark got an extra publicity push due to a medical misunderstanding.

It is thought that malaria was native to the UK through to the end of the 19th century – and as world trade grew, malaria was also more common among travellers to tropical destinations. Historically referred to as the 'ague', there were major outbreaks in particularly warm spells, but ague seems to have declined in Britain through the 19th century, in part because of drainage of swamps and in part because of the use of an extract from the bark of the cinchona tree, sometimes known as Peruvian bark.

Cinchona bark was known as *quina* by the Inca locals who first used it as a treatment, giving us the derivation of the medical drug used for malarial treatment all the way through to the 1940s – quinine. At the strength required to counter the impact of malaria this has significant side effects that have led to its replacement, though in very dilute form it is still found in tonic water. However, quinine doesn't have the same biomimetic quality as aspirin, because the refined drug is not just inspired by the active chemical in the natural source – quinine *is* that same substance. It is a useful comparator for showing how aspirin is a true biomimetic product.

Cinchona bark was expensive, so some effort was put into searching for a cheaper substitute. The misunderstanding that led to an increase in use of willow bark was from an improvement in malaria sufferers when treated with it. In reality, the willow was only suppressing symptoms of malarial fever, but it was thought that, like cinchona, willow bark could control the impact of malaria overall. A report for the Royal Society recommending the use of willow bark for ague meant that what was already a useful painkiller found a significantly wider market.

Like many medical treatments of the period, willow bark extract provided a mixed blessing. This was a time when, for example, the extremely dangerous element mercury was used for treating syphilis. Unless used carefully, the cure could prove fatal faster than the illness. Similarly, willow bark carried its own dangerous side effect. The active chemical in the bark is a salt of salicylic acid. This chemical compound, named after the willow tree (salix) can induce intense stomach pains and can even cause internal bleeding. Pain might be reduced in one part of the body, only to create an even worse condition elsewhere. However, the need for a painkiller and anti-inflammatory was often worth taking the risk.

It was (in part) dealing with the side effect issue that resulted in the development of one of the first wonder drugs – a drug that was biomimetic as it copied the active approach of salicylic acid but was modified to make it safer to use. In 1899, the German chemical giant Bayer brought out a new product called Aspirin, a trade name for a compound called acetylsalicylic acid. This was first produced by the French chemist Charles Gerhardt in 1853. He made use of acetyl chloride – effectively acetic acid (vinegar), replacing the oxygen-plus-hydrogen (OH) part of the molecule by reacting it with the sodium salt of salicylic acid. Aspirin stops the enzyme cyclooxygenase from working, which results in reduced production of hormones that both transmit pain messages to the brain and induce inflammation.

Although acetylsalicylic acid can still cause damage to the stomach lining, it is significantly less aggressive than salicylic acid from willow or meadowsweet but is just as good at reducing pain and inflammation. It mimics the positive action of the natural product while reducing its unwanted side effects. Bayer came up with the name 'Aspirin' from a shortened version of the old German name of the compound, *acetylspirsäure*, tacking on the 'in' ending to bring it into line with Bayer's then popular cough suppressant medicine and other leading brand-name product, Heroin. Others would make use of products based on acetylsalicylic acid, but only Bayer could call their painkiller Aspirin.

The reason we in the UK can now refer to aspirin with a small 'a' is, bizarrely, a result of the signing of the Treaty of Versailles on 28 June 1919. This document specified the formal reparations that would be laid on Germany at the end of the First World War – and is now seen by some as so punitive that it was one of the reasons that the Second World War occurred. Inevitably, much of the treaty concerned itself with the delineation of boundaries and imposing military restrictions on the losers of the war. But the document also dealt with finance and trade, the aspect that had a particular impact on German development in the interwar years. This involved huge payments and a requirement to hand over steel and coal, but also, emphasising what a wonder drug aspirin was considered to be, explicitly gave the signatories of the treaty the ability to freely use the name 'aspirin'.

In Germany, Aspirin (with a capital A) remained a trademark of the Bayer company: it still is today, not only there but also in many other countries around the world. However, in the UK and other signatory countries, the treaty meant that the name aspirin could be freely used. The primary influence on the appearance of the name of a medication in a treaty to end a war seems to have been the last pandemic to hit the world before Covid-19. Spanish flu* was a devastating

---

* The name 'Spanish flu' seems to have been the result of wartime news control, as reports of outbreaks in neutral Spain were not suppressed as much as those from elsewhere.

global pandemic, estimated to have infected around a third of the world's population and to have killed between 20 and 50 million people. Aspirin was a key medication used to moderate the fevers called by influenza and was tied up in the public mind with the pandemic that coincided with the end of the First World War.

All the way through to the 1970s, aspirin's dominance was unchallenged, but it has now been largely supplanted by the more stomach-friendly paracetamol. This is called acetaminophen in the United States, better known under trade names like Panadol (owned by GlaxoSmithKline) and Tylenol (owned by Johnson & Johnson). Paracetamol has also been joined by so-called 'non-steroidal anti-inflammatories' such as ibuprofen and diclofenac as popular painkillers. However, the relatively short-term drop in aspirin's fortunes was countered a decade later when there was increasing evidence that aspirin could help prevent heart attacks and strokes by reducing the tendency of platelets – fragments of blood cells – to clump together and obstruct blood vessels.

For whole generations, aspirin was the first line of attack when suffering pain, and now it has revived its fortunes thanks to its action on platelets. Around 35,000 tonnes of aspirin are consumed each year. Few other medical compounds can boast the staying power of the humble drug that featured in the Treaty of Versailles and that mimics the action of willow bark in a safer fashion.

Taking on sugar

Historically, the concept of a biomimetic product inspired by a foodstuff would seem ridiculous. Food was inherently based on natural materials. And that is often the best thing for us – there is no doubt that many modern foodstuffs are too processed to retain their natural benefits. However, just because something is natural does not make it healthy. Some of the most potent poisons, such as ricin or the botulinus toxin used in botox, are natural. And it's easy enough in a modern environment of plenty to over-consume some natural substances.

So, for example, the very common natural chemicals known as sugars make our food taste better – but with the increasing realisation of the health impact that comes with the over-consumption of sugars, we have seen a huge rise in the development of sweeteners than mimic the natural effect without their dietary downside. Although these are inevitably biomimetic in use, artificial sweeteners have something of an accidental history, notably in the first big name, saccharin.

If we call something 'saccharine' it is not a compliment. The word suggests something that is cloyingly sweet, or metaphorically far too goody-goody. It's downright sickly and unattractive. The word was in use as early as 1674 to mean sugary, taken from the medieval Latin term for sugar, *saccharum*. Yet this was the name given to the first of the artificial sweeteners. It's hard to imagine a modern branding agency would have allowed such a slip-up: the chemical compound might have dropped the final E to become saccharin, but it's far too close for comfort.

The substance in question, more formally known as benzoic sulfilimine, with the chemical formula $C_7H_5NO_3S$ was granted a patent under the name 'saccharin' in 1884 for Russian chemist Constantin Fahlberg. There is considerable doubt over who made the actual discovery of the substance as a sweetener, though there is agreement that it was an accident. According to Fahlberg's account, in 1879 he had been working on compounds derived from coal tar at Johns Hopkins University in Baltimore. When he got home (and presumably failed to thoroughly wash his hands) he sat down to eat with his wife.

When he took a bite from a bread roll, Fahlberg noticed it seemed unusually sweet before developing an aftertaste of bitterness. His wife noticed nothing odd about the taste of her roll – but licking his hand also gave Fahlberg a sweet-then-bitter taste sensation. In an approach that would definitely not be recommended now, Fahlberg went back to the lab and tasted the different chemicals he had been working with, narrowing the substance down to benzoic sulfilimine.

This all sounds entirely straightforward, had it not been the case that Fahlberg's boss, American chemistry professor Ira Remsen, also claimed to have had the bread roll experience and to have discovered saccharin. The pair jointly published on the substance, but it was Fahlberg who obtained the patent, cutting out his old laboratory head. Remsen claimed he wasn't worried about the financial gains that Fahlberg made, but that he should have received credit, as would be expected as head of the lab. By the time Fahlberg obtained his finger-licking-good patent, he had left the university. Remsen is said to have commented: 'Fahlberg is a scoundrel. It nauseates me to hear my name mentioned in the same breath as him.'

Just as the First World War would transform the fortunes of aspirin, it was the war's impact on sugar production that resulted in saccharin being used as a popular substitute that stayed in use when the war ended and sugar supplies bounced back. However, it would not be until the 1960s and 1970s, when the idea of counting calories was taking hold in Western countries increasingly obsessed with diet, that saccharin reached the heights of its popularity. Because it sweetens with zero calories, it seemed an excellent biomimetic alternative to sugar, although that bitter aftertaste that Fahlberg or Remsen supposedly noticed has always made it second best to the real thing for taste. Little pink packets with brand names like Sweet'N'Low began appearing on café tables alongside the sugar.

Challenging sweetness

Saccharin hasn't entirely disappeared – you'll still find those pink packets, and it is still used in food manufacturing – but reflecting the reality that it has been one of the most widely used biomimetic products, it has faced a number of legal and medical challenges along the way. It was investigated in the United States in 1907, 23 years after it was first patented, because sugar producers considered it a threat – before First World War shortages made it more of a necessity, it was considered a low-price substitute that devalued existing sugar-based products. Since then, along with most artificial sweeteners, it has been responsible for a number of health scares. Saccharin has been accused of causing cancer, but the only evidence was based on experiments with rats (usually consuming impossibly large amounts of the compound), and there have been no equivalent discoveries of problems facing primates like humans. Some countries did ban saccharin for a while but have now largely restored the compound as a safe product for use in foodstuffs.

The biomimetic action of saccharin (and other artificial sweeteners) is in mimicking the ability of sugars to trigger the 'sweet' sensation in the tastebuds of our tongues. According to a still widely believed myth, tongues have different areas corresponding to taste sensations such as sweetness, saltiness and sourness. I remember doing an experiment at primary school where we prodded our tongues and failed to do what our teacher hoped we would do – detect different taste sensations as a result of activating the relevant tastebuds. But had we succeeded it would have been a placebo effect – because the tongue simply isn't broken down into areas that deal with different tastes.

The sensation of taste is the result of receptors in the tastebuds responding to different chemical substances dissolved by our saliva. Some of these receptors are relatively simple. The sensation of saltiness, for example, simply depends on the presence of sodium ions, most typically from dissolved sodium chloride. But sweetness is a more complex sensation, combining input from multiple different receptor sites. And it's this that saccharin and other sweeteners do well.

The benefits of saccharin were originally cheapness and easy availability, but they are now more about health improvement. Artificial sweeteners mean that those suffering from diabetes can still satisfy their natural craving for sweet food – which seems to be an ancient response to both detect high energy content in foods and to notify us of an absence of toxins, which are often bitter. For those looking to lose weight, or simply keep sugar intake down, the idea of sweetness without calories will always be popular.

These days, we are less likely to see saccharin in coffee shops, as it has largely been replaced by aspartame. This more complex sugar substitute, sold under brand names such as Canderel and NutraSweet, is closer to sugar in taste with less of the bitter aftertaste. Discovered back in 1965, this was another accidental discovery, in this case by James Schlatter of the American Searle company (now part of Pfizer). Like Fahlberg (or Remsen), Schlatter discovered the sweetness of the compound (he was working on it as an anti-ulcer drug) when he licked his finger to turn a page.*

As with saccharin, there was concern over aspartame when rats that were fed huge amounts of the substance developed brain tumours, and it has been subject to regular anecdotal suggestions of causing ill effects, but aspartame has become one of the most safety-tested substances in history and passes with flying colours – unlike many natural substances. A good example is coffee, which contains around 1,000 chemicals, of which fewer than 50 have been given detailed safety testing. Some of these will almost certainly be carcinogenic in large quantities, but they are entirely safe in the amount present in a cup of coffee. Compared, for example, to the more recently developed alternative natural (and hence not biomimetic) sweetener Stevia, based on an extract of the *Stevia rebaudiana* plant, we are far more certain of the safety of aspartame.

Saccharin does still creep into the mix sometimes as it has a longer shelf life than aspartame, so is sometimes included to keep a product sweet as aspartame loses its impact. Taken together it's arguable that saccharin and aspartame are the most widely encountered of all biomimetic

---

* Arguably chemists should know better than to lick their fingers in the workplace, and we don't have a record of how many have done so and made themselves ill or worse.

products. They are certainly better known than another, very different biomimetic compound that sounds more suited to the armoury of a bad guy in a superhero movie than a substance we encounter in everyday life.

A bitter pill

Where nature is generously provided with sources of sweetness, often present to tempt animals to consume a fruit, thus ingesting and later spreading seeds, evolution has provided many other plants with chemical defences to *prevent* them from being eaten, often making the plant bitter to human (and other animal) taste buds. The idea of warning someone off with a bitter taste has been expanded in a biomimetic way by finding a substance that beats nature at its own game.

Before science fiction reached the modern age (this is still sometimes the case in superhero fantasies) there was a tendency to devise all kinds of weird and wonderful new substances that had nowhere to sit on the periodic table. At the time, this reflected the way that new elements were being produced in the laboratory, especially those like radium that had near-magical powers. Often, the fictional elements or alloys would be incredibly hard (typically used for armour of spaceships) or would have supernatural-seeming powers (like radium, they also had a tendency to glow in the dark) – think, for instance, of Superman's kryptonite, which was originally described as an element unknown on the Earth, though later stories described it as an alloy. If there is one real substance in biomimetic chemistry that sounds as if it belongs on the fictional lab shelves alongside kryptonite, adamantium, feminum and all their cousins, it is denatonium – but this one is real, if not an element.

Denatonium is the snappy short name given to a substance that chemists prefer to call phenylmethyl-[2-[(2,6-dimethylphenyl)amino]-2-oxoethyl]-diethylammonium. This is an aromatic organic compound – aromatic meaning that it has so-called benzine rings of six carbon atoms in its structure, and organic meaning simply that it contains carbon. In practice, denatonium is used as a benzoate – a salt of benzoic acid, which is an acid based on a single benzene ring. Denatonium benzoate's claim to fame is in its ability to more than mimic those nasty-tasting plants – it is the most bitter substance yet to be discovered.

This unreactive, colourless, odourless compound was first produced accidentally in 1958* by Scottish pharmaceutical manufacturer T&H Smith, later Macfarlan Smith. This time, the intended product was an alternative to an anaesthetic used by dentists called lignocaine. Denatonium benzoate turned out to be no help in making fillings painless, but it was soon discovered that just a few parts per million were enough for this aggressively unpleasantly flavoured compound to render a substance it was added to distasteful to humans.

Bitterness is one of the five basic tastes alongside the sweetness we've already met, saltiness, sourness and the late addition of the savoury taste umami.** As we have seen, it is a myth that the tongue has receptors for different tastes in specific regions of its surface, but our tongues *are* more

---

* Accidental discovery is something of a pattern in biomimetic chemical compounds.
** Umami should surely have been called savoury, alongside using sharpness instead of bitterness, to make the key flavours the five S's.

sensitive to some tastes than others – and bitterness is the one that grabs our attention most strongly.

It seems likely that this is because many toxic substances contain chemicals that stimulate the bitter taste reaction in the tastebuds, though perversely, some mildly toxic substances that many of us enjoy – coffee, hops in beer and the quinine in tonic water, for instance – do have this bitter kick. Usually these are tastes that we acquire over years of exposure: children tend to have an aversion to them, but as we get older our brains can override the palate, enabling us to appreciate more sophisticated taste combinations.

Like sweetness, bitterness is not the responsibility of a single receptor on the tongue, nor is it a response to a simple presence like the sodium ions that produce saltiness. A collection of genes that encode for a total of 25 different taste receptors, each reacting to a different class of compound, combine to produce the impact we label bitterness. We don't even all respond to every single bitter trigger. For example, around a quarter of the population can't taste the compound propylthiouracil (often known more simply as PROP), which is similar to the compounds producing the bitter flavours of cabbage and the quinine in tonic water. The ability to detect this taste (and this kind of bitterness) is dependent on a single gene, TAS2R38. A single letter variation in that gene – a so-called single nucleotide polymorphism or SNP – is responsible for whether or not a person can experience this taste sensation.

By contrast, reaction to denatonium benzoate is more or less universal – and that reaction is overwhelming. To put it into context, quinine, the archetypal TAS2R38 bitter substance, is used as the benchmark for bitterness, and for those without the gene variation, the tongue can pick up the bitter kick of quinine at a concentration of around 0.000008 moles per litre. That's pretty dilute – compare it with the salt in seawater, which is typically around 0.6 moles per litre. By contrast, denatonium benzoate requires a thousandth of the concentration of quinine to be detected.

Given that denatonium benzoate does not exist in nature, our ability to detect such small quantities does not reflect a natural response to it – it is more likely an accidental response to the way that taste receptors on the tongue react to this particular compound. However, this is an accident that has found plenty of practical use. Denatonium benzoate is marketed under brand names such as Bitrex, BITTER+PLUS and Aversion, which are all substances known as bitterants or aversive agents.

The idea is simple – if you have a product that could be consumed but shouldn't be, you add some denatonium benzoate and even small quantities of it will put people off. As many as 30,000 children are taken to hospital each year in the UK with suspected poisoning. Usefully, children seem particularly sensitive to the bitter taste of denatonium and are repulsed by it, making it a helpful safety product.

One good example of a substance that cries out for added bitterness is the otherwise sweet-tasting but poisonous anti-freeze ethylene glycol. Denatonium benzoate is also used in some rat poisons (as it happens, rat tongues are a lot less sensitive to the compound than human tongues). The substance is also used to 'denature' ethyl alcohol, the compound found in alcoholic drinks, making it undrinkable. This means that when ethyl alcohol is used for cleaning or as fuel it can be sold without the large tax burden that usually accompanies booze – a particularly common requirement with the increasing use of bioethanol.

Perhaps the most direct application of denatonium benzoate is to help those who can't stop themselves from biting their nails. A mixture containing a small amount of the compound is painted onto the fingernails and when the sufferer attempts to chew them – often doing so unconsciously – the bitter taste quickly alerts them to stop.

Taste is one of the weaker senses – foodies will tell you that the sense of smell is more important, and taste is a sense that we rate lower than sight or hearing, but it is still important as demonstrated by the range of biomimetic chemicals we use to influence our tastebuds. But biomimetics in foodstuffs can go considerably further in mimicking the experience of eating food usually taken from sources we wish to avoid. Whether it's vegetarians tucking into a fake burger or vegans calling 'cheese' something that has no relationship to milk, there is now a whole industry in faking food.

Making the meat

In my freezer, I have some quarter-pound burgers from a well-known British food company named after a pop star's late wife. They are tasty burgers – not, admittedly like a homemade special, but perfectly acceptable in comparison to other shop-bought frozen burgers. But they have had nothing to do with a cow: they are entirely vegetarian in origin.

Looking at the contents, the burgers are made up of 58 per cent soya protein, oil, water, onion, chickpea flour and a range of flavourings*. They provide the eating experience of

---

* Other vegetarian burgers will have variants on these ingredients, but those listed are representative.

taste and texture that we expect from a burger, but without an animal being harmed in the process. Although there are a few dubious-sounding items in the contents list (methyl cellulose stabiliser and unspecified 'flavourings'), they seem fairly harmless, though we can't doubt that the product is highly processed.

Mimicking a meat product requires a whole range of sensory deception. Most vegetarian food makes no attempt to pretend it is meat-based – and that is fine. I enjoy a spicy beanburger in its own way as much as a meat-like veggie burger. But there are times, particularly if like me you aren't a vegetarian but want to eat less meat both for dietary and environmental reasons, when getting the look, taste and texture of the real thing makes all the difference.

Some attempts at biomimetic meat go even further. There are plant-based burgers, for example, that ooze fake blood – but many would argue that this is an unnecessary step. And there is also the possibility of lab-grown meat. Here we are dealing not with vegetables pretending to be meat but actual animal protein that is grown as a cell culture.

Lab-grown meat gets us over the issues surrounding traditional meat production but provides what is to all intents and purposes the real thing, so while in a sense biomimetic, it isn't in the usual meaning of imitating the function of a natural substance but using different materials. Having said that, there are still issues that have to be overcome to get the texture, taste and appearance of lab-grown meat right. This is why many early attempts have tended to focus on burgers, where the minced texture is easier to achieve than the complex structure of, say, a fillet steak. Similarly, the cost of these lab-grown cell cultures was initially prohibitive, though it will no doubt become more affordable with time.

Cheesed off

For vegans, it's not enough to stop eating animals: even eggs and dairy are off limits. While it's significantly harder to justify taking this extra step, vegan food has become popular. (In fact, many vegetarians complain they can't get a decent vegetarian meal in a restaurant any more because vegan food has taken over.) One of the biggest challenges has been in producing vegan cheese that has decent flavour and texture.

Here what is being mimicked is a product made from a natural substance – milk. But there is added difficulty in the complex nature of cheese, especially if you look at something like a mature cheddar, where part of its essential nature is the lactose crystals that form as it ages. While vegan 'cheeses' have got better, many vegans will admit that they are not ideal.*

A significant problem here is that even more so than the vegetarian burgers, a vegan cheese is a highly processed product. If someone's reason for going vegan is health-based, then it's important to be aware that the chances are that the vegan alternative will be significantly more processed than the natural original. Looking at the contents lists of a range of vegan cheese products, there are usually mostly natural ingredients (water, coconut oil, potato and maize starch, guar gum and yeast extract are common) – but the fact remains that these are industrial products. That guar gum, for example, used here as a thickener, does come from guar beans,

---

* Vegan cheeses are, nonetheless better than the alternative. I remember going for a pizza with my then literary agent who was a vegan before it was trendy and watching with amusement his attempts to persuade the server that it was possible to make a pizza without cheese.

but will have been added to the machinery producing the 'cheese' as a greyish powder.

The risk with consuming highly processed foods is that they rarely provide as balanced and effective a nutritional package as a natural foodstuff. Admittedly, cheese is primarily fat, and as such was never going to be wholly healthy in the first place, but just as we have now discovered that butter is probably better for you than a lower fat but highly processed spread, so it seems likely that vegetarian 'cheese' is not as nutritionally beneficial as the real thing.*

Despite any concerns, though, vegan and vegetarian food is on the rise – and plenty of these products are biomimetic in providing more or less accurate duplicates of natural foodstuffs with the aim of giving the eater the same pleasure with less impact on the environment or a more healthy meal. A good example is the range of alternatives to cow's milk, whether it be products derived from soya, oats, almonds or others. Again, these are more processed than conventional milks (and in some cases sweetened). However, the alternatives generally have significantly less impact on our climate, though in some cases, particularly that of almond milk, there are secondary environmental impacts involved in their creation. These include the requirement for large amounts of water in parts of the world where rainfall is scarce, as well as heavy pesticide use and replacing food crops or grazing land.

In producing a good food replacement, texture is hugely important. I am sure I am not alone in that the majority of foods I don't like, I dislike because of the texture (think, for

---

* I am assuming here we are dealing with a real cheese, rather than an abomination such as spray cheese.

example, the sliminess of avocado or of smoked salmon). And texture is all a matter of the structure of the foodstuff. This makes a useful bridge from the previous chapter – when we were considering materials – to the next, where structures become all-important in their own right.

STRUCTURES, SHAPES AND NETWORKS 5

Nature is home to a wide range of structures, from the simple but powerful form of a crystal to the elegant lattices of a honeycomb. Not surprisingly, these have given practical inspiration to designers, engineers and architects.

From mound to megastructure

It is not uncommon to take architectural inspiration for the appearance of a building from nature, whether it's fitting out a tower block with a roof garden or the forest-inspired styling of Gaudi's Sagrada Família basilica in Barcelona. However, biomimetics requires an inspiration based on function rather than a mere visual starting point. Learning a lesson from nature in building design is distinctly less common. Perhaps the most surprising such inspiration, given architects' usual enthusiasm for style, is the ugly, externally formless structure of the termite mound.

Because they don't have such an obvious benefit to humanity compared with the pollination and honey production of bees, termites rarely get the same good press. Many termite species eat wood and can cause unwanted damage to wooden buildings and to food crops, while their mounds are both unsightly and can end up providing obstacles in unwanted locations, such as dirt runways in rural airports. A relation of the understandably despised cockroach, the termite does not provide an obvious source for an architectural vision.

But within their nests, termites have a complex superorganism existence, which rivals that of some species of bees or ants. In a superorganism (a species that is scientifically referred to as being eusocial), the true organism is the whole colony rather than the individual insect or mammal.* The individual insects act more like the different kinds of cells or organs in a body than as true organisms with an independent existence. This is made possible by the specialisation of individuals – like bees, termites have workers, soldiers, fertile males (known as kings) and at least one queen. The colony works together as a unit.

One capability that arises from this eusocial behaviour is the construction of complex structures. Many of us will be more familiar with the surprisingly sophisticated structure of a wasp's nest or beehive, but by comparison with termite mounds these are modest. Not all termite species build mounds, but some create large, intricate earthworks to live in. The most impressive mounds can be as much as 9 metres (29 feet) tall – these are the insects that can best

---

* Most eusocial animals are insects, but the distinctly weird naked mole rat is a eusocial mammal. (And, yes, insects are technically animals.)

be described as architects. However, there is more to some termite mounds than an impressively large structure. These mounds are effectively air conditioned.

Natural cooling

Termite activity produces a considerable amount of heat inside the mound. There can be more than a million insects living in a single structure – not surprising when you consider that one colony typically shifts around a quarter of a tonne (550 pounds) of soil in a year, building and maintaining their nest. All this effort generates hot air. The internal structure of the mound allows the heated, rising air to move upwards. In some species' mounds, the air then exits through a chimney at the top, sucking in cooler air from ground level, while in others, the air circulates internally, passing through underground chambers that naturally stay cooler than the mound above. This means that the temperature doesn't rise too high and stays more constant than it otherwise would. This is important for some termite species, as their mounds are not just homes, they are farms.

These agriculturally inclined insects encourage the growth of fungus in their mounds – but the fungi they specialise in tend to be temperature sensitive. Keeping the temperature relatively constant makes for optimal growth. The architectural 'design' – we always have to remember that evolution is not about having a conscious direction but more about keeping the results of beneficial accidents, but it is a difficult word to avoid – is as much about making the environment work for the fungus as for the insects.

Other termites, notably the so-called compass termite (these are Australian termites from the species *Amitermes meridionalis* and *Amitermes laurensis*) build mounds shaped rather like tombstones (though narrowing towards the top) that are roughly lined up in a north-south direction. This gives them even better temperature control, as the mounds heat up quickly in the morning sun but get minimal exposure in the midday peak where they present a restricted profile. As the mound is heated by the morning sun, it produces the effect we have already met, known as a solar chimney. Air inside the mound rises, warmed by the sun, sucking in cooler air from below. Such chimneys had been used in buildings in the Middle East and some parts of Europe for centuries before there was an understanding of termite mounds, but now architects are consciously imitating termite building styles.

A dramatic example of a termite-inspired concept is the Eastgate Centre in the Zimbabwean capital of Harare, designed by local architect Mick Pearce. Many modern office blocks make use of swathes of glass, which results in over-heating, requiring expensive, high-power air conditioning to keep the building cool. But the Eastgate Centre's structure, which has recessed windows and large amounts of exposed material that is good at absorbing heat, intentionally reduces the opportunities for direct sunlight to warm up the building.

At the same time, a total of 48 chimneys are used to provide a termite mound-like cooling effect. This is aided by a large basement space that stores cool air at night which is then pulled through the building as the solar chimney effect kicks in. The result is a 35 per cent reduction in energy use compared to the expected air conditioning load required by a similar sized conventional build. Although the system works best when using fans to drive the process at some times of day, it still functions during the relatively frequent power outages in the city, unlike conventional air conditioning.

This is a wonderful story – but as so often is the case with biomimetics, things aren't as simple as they appear. This is a case of one of the most intriguing types of invention – invention by misunderstanding. This can happen when an inventor thinks they know how something works and designs something accordingly, only to find out that their initial information was incorrect. This has resulted, for example, in the invention of a better electron microscope than the first model, when scientists made a mistaken assumption of how it could possibly work. They were wrong, but their attempt at reverse-engineering was better than their inspiration.

The assumed mechanism of the termite mound brought Mick Pearce to a more sophisticated version of the solar chimney. However, recent research suggests that termite mounds don't gain much if any of their temperature control by managing air flow. In experiments undertaken in 2008, it was found that closing the 'chimneys' at the top of mounds to stop the flow of air had hardly any impact on the temperature inside. Instead, the relative constancy (which proved significantly less than had been thought) was down to the large amount of the mound that was underground. It was this thermal sink of air that stayed at a roughly constant temperature that provided the regulation.

Was the building design still biomimetic? Certainly, the inspiration was from nature, even if the mechanism was partially misunderstood. It's arguable that in architecture, the intention is enough. But as is often the case with biomimetics, Pearce's work is still a rare outlier. The vast majority of office buildings, for example, still have swathes of glass and rely on energy-hungry air conditioning systems. Perhaps as climate change becomes an even more pressing issue, passive cooling systems based more on natural mechanisms will become more common.

From crystal structures to the bubbles of Eden

The inspiration that termites provided was all about internal mechanisms – it's unlikely anyone would want a building with an external structure that was based on a termite mound. But a pair of other examples demonstrate different ways that ideas could come from nature without the exact mechanisms being duplicated. These are Joseph Paxton's 1851 Crystal Palace and Nicholas Grimshaw's Eden Project biomes in Cornwall built in 2001. Paxton was both a horticulturalist and a building designer. He was the first person in the UK to cultivate the *Victoria amazonica*, the largest of the water lily family, which has a very clear box-like rib structure on the underside of leaves that can be up to 3 metres (10 feet) in diameter.

Paxton would later say of these lily leaves that 'you will see that nature has provided it with longitudinal and transverse girders and supports, on the same principle that I, borrowing from it, have adopted in this building'. The building he referred to was a glasshouse he designed for the Duke of Devonshire's Chatsworth estate, where Paxton had become head gardener at the age of twenty. This structure, known as the great conservatory, was built to house these lilies (then known as *Victoria regia*) as well as being inspired by them. Paxton is said to have tested the strength of the lily pads by putting his daughter Annie on one, though this sounds more like a publicity stunt than anything practical.

Work started on the conservatory in 1836 and it had impressive proportions – 69 metres (227 feet) long by 37 metres (123 feet) wide – the largest glass structure of its time with curved glass panels supported on those lily-inspired ribs that still gives it a modern look today. These same principles would be carried even further in his far more challenging Crystal Palace, built for London's Great Exhibition of 1851. This structure was massive at 563 metres (1,848 feet) long, 124 metres (408 feet) wide and 33 metres tall (108 feet) and proved a huge success with the public. Arguably, Paxton's approach was to be continued, at least as an inspiration if not in practical detail, in many of the metal and glass structures that are built to this day.

Moving forward 150 years, the unsupported structures of the bubble-like Eden Project biomes are geodesic domes,* a style popularised by American engineer and architect Richard Buckminster Fuller. But there was an extra twist to the design of the Eden Project biomes.** The location of the structures is at the bottom of a disused china clay pit, which was in its final days of operation when the biomes were designed. The lack of a finished site was a nightmare for the architects, as it was not clear what levels each of the connected domes that makes up each biome would be on. Here, the biomimetic inspiration was a soap film. These ultra-thin layers that make up the surface of a soap bubble

---

* 'Geodesic' structures are three-dimensional forms where the faces are made up of triangles.
** A biome is a geographical area with distinctive biological flora and fauna. At the Eden Project the two biomes are effectively massive greenhouses, one with a rainforest environment and the other a Mediterranean feel.

act as natural calculators, automatically devising the solution to an architectural problem.

In a number of circumstances, the laws of physics enable an object to carry out what is effectively a mathematical calculation. One such example is the catenary – the shape that a rope, cable or chain will make when it is suspended between two points. That shape is the solution to a mathematical function called the hyperbolic cosine. The inverse square nature of gravitational attraction ensures that (within the limits of flexibility of the material) the shape produced reflects that function.

Similarly, a soap film takes on the shape that will minimise the surface area of the film required to enclose a particular volume. It automatically solves this problem. One consequence of this ability is that when two soap bubbles join, they naturally do so perpendicular to each other. Although the ethylene tetrafluoroethylene material used in constructing the Eden Project domes is not as flexible as a soap film, it is close enough to soap film behaviour to ensure that the self-supporting structures erected on an unpredictable and potentially changing surface were still able to function effectively without having to be rebuilt.

In both Paxton's work and at the Eden Project, a natural structure inspired an artificial one. However, in at least one biomimetic example, a field of mathematics inspired by nature then led on to a different kind of functional design.

Fractal inspiration

The mathematical concept in question was that of fractals. The fractal was originally dreamed up by the French–American mathematician Benoit Mandelbrot while thinking about an apparently simple question about the island of Britain. What is the distance around the British coast? As Mandelbrot realised, this is a near-impossible question to answer. If you measure the distance with a metre rule or yardstick you will get one distance. But if you use a shorter measure, the distance will be further, as you will be able to get into more of the cracks and crannies around the coastline. Depending on how it is measured, the distance around Britain's coast has been given as anything between 1,700 miles (2,700 km) and 11,500 miles (18,500 km). None of these answers can be taken as 'the right answer'.

The approach Mandelbrot would take in dealing with this problem he described as follows: 'I conceived and developed a new geometry of nature and implemented its use in a number of diverse fields. It describes many of the irregular and fragmented patterns around us, and leads to full-fledged theories, by identifying a family of shapes I call fractals.' Fractals, then, were inspired by nature – but they have also come through into the real world of structures. A fractal is a shape that is usually self-similar – so zooming in on a smaller part of it resembles the larger whole – and fractals have the mind-boggling mathematical formulation of having fractional rather than integer dimensions (hence the name 'fractals').* Many natural forms are roughly fractal, from trees to mountain ranges and clouds.

Fractals have been used widely in computer graphics – for example, in constructing artificial landscapes for movies – and were briefly used as a mechanism for compressing

---

* The idea of fractional dimensions can seem unlikely, but it uses a different concept of what a dimension is to our usual simplistic view – for some shapes (for example, squares and cubes), the minimum number of smaller copies of the shape that can fill the shape is two to the power of the number of dimensions.

photographs to take up less disk space, but in 2012, Satoshi Sakai, a professor at Japan's Kyoto University, came up with the idea of a biomimetic fractal sunshade. This was inspired by discovering that tree leaves have a lower temperature than the surrounding air when exposed to sunlight. It seems that the fractal structure of a tree's canopy provides a more effective sunshade than a straightforward solid barrier. Sakai devised the Comolevi Forest Canopy (the name 'Comolevi' is based on the Japanese term for sunlight filtering through leaves).

Sakai could have based his artificial canopy directly on the shapes of tree leaves, but instead he made use of a simpler fractal design based on a mathematical structure known as a Sierpiński gasket or triangle.

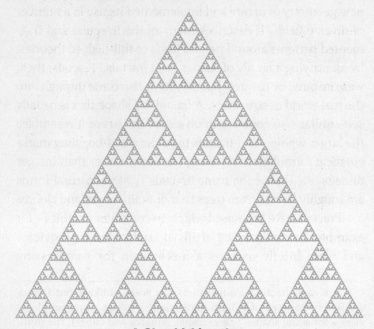

A Sierpiński gasket.

By Beojan Stanislaus, CC BY-SA 3.0

The basic gasket takes the form of a solid equilateral triangle with interior segments removed. The first cut out is the biggest equilateral triangle the opposite way up that can be removed while leaving the material intact as a single piece. Then, each of the three remaining smaller triangles has the same process applied to it, and so on. Although the gasket never runs out of material, its surface area tends to zero as more and more triangles are removed. The gasket has a fractal dimension of around 1.59.

Sakai's finished product, made from canvas, combined two different three-dimensional gasket designs, one above the other (the lower one is based on hexagons rather than triangles) reflecting the difference in size and influence between the higher and lower leaves on a tree. The result was significantly more effective at cooling the area below than a continuous piece of canvas. The gaskets provide shade but prevent the canopy itself from heating up as much as a solid canvas by allowing air to pass through it. By comparison, with a conventional canopy, a build-up of warm air beneath, frustrated in its attempt to rise, produces an unwanted hotspot. In tests, the canopy produces up to a 12 °C (21.6 °F) cooler space than does a continuous canvas canopy.

This effect has plenty of possible applications for managing the temperature of outside spaces and it has even been suggested that buildings located beneath such a canopy could be cooled this way. This has the potential to make a significant contribution to reducing air-conditioning energy use – which would be a boon as climate change progresses. Although on sale, manufactured by the Japanese company Losfee, and installed in a number of locations, as yet the product has not been widely adopted.

Functional shapes

Although strictly speaking the inner configuration of a structure and its overall shape are different things, in practice it is often near impossible to draw a line between the two. However, there are circumstances where an external shape can be divorced from the structure that supports it. We often think of the shape of products as being primarily about aesthetics – there is no doubt that some shapes look better to our eyes than others. Yet shape can also have a direct impact on function. And in a small number of cases, biomimetics have provided the reason for the adoption of that shape.

Streamlining is, perhaps, the clearest example of a shape's potential to have functional impact. Historically, the design of railway engines probably gives us the best illustration of the way that form and function can interact. Even back in the days of steam, some locomotive designers thought about the impact of streamlining. The world record for steam engine speed is held to this day by *Mallard*, Sir Nigel Gresley's A4 class locomotive built for the London and North Eastern Railway in the 1930s.

It was the elegant, streamlined form of the A4 locomotives, among the first to be tested in a wind tunnel, that enabled *Mallard* to reach a speed of 203 kilometres (126 miles) per hour on 3 July 1938. The majority of steam locomotives of the day had little if any streamlining – it was arguably an unnecessary complication when trains were not expected to reach high speeds. Yet today, when *Mallard*'s record is commonly exceeded by scheduled high-speed intercity trains, streamlining is a standard part of the design process – and in the need to take streamlined designs to the next level, engineers have made innovations that have been driven by biomimetics.

Even by current streamlining standards, the Japanese Shinkansen bullet trains are unusual with their dramatically long, beak-like fronts. Where typical high-speed trains, such as the French *Train à Grande Vitesse*, operate in the 200–300 kilometres (125–186 miles) per hour range, Shinkansen push up a little further to 240–320 kilometres (150–200 miles) per hour – hardly a huge increase in speed. But in making the impact of high speeds on the environment more acceptable, those extended noses have a specific role, which is modelled on a natural exemplar – the kingfisher.

These gem-like river birds take on a dart shape when diving into the water to catch a fish. This shape makes the transition from flying through the air to the medium of water, which is around 800 times denser, far slicker. It leaves the bird in control as it arrows in on its prey. Although high-speed

**The nose of a Shinkansen train, influenced
by the diving kingfisher.**
MaedaAkihiko, CC BY-SA 4.0

trains don't need to make the transition from air to water, they do enter and leave tunnels and cuttings, which presented a particular problem for the Shinkansen's designers.

On Japan's crowded islands it was not always possible to route high-speed lines very far from housing, while the mountainous landscape makes for many tunnels. But when a train travelling at high speed enters a confined space, it produces a dramatic change in the density of the air ahead of it. Because the train's profile takes up a considerable amount of the cross-section of the tunnel, it acts like a piston in a cylinder, compressing the air in front of it, which produces a pressure wave. When the train then leaves the tunnel (or to a lesser extent a cutting) the result is a loud bang as the compressed air expands in all directions, an effect not dissimilar to a sonic boom.

Encouraged by a bird enthusiast in their management, the team working on the development of Shinkansen studied the kingfisher's head and beak shape and its fluid dynamic properties. The resultant beak on the trains not only reduced noise as a result of a 30 per cent reduction in air resistance, it also pulled down energy consumption by 13 per cent. With a better ability to cut through the air, the shape provides less resistance as the train powers down the track. The potential next generation of Shinkansen, the ultra-high-speed L zero magnetic levitation trains that are intended to run at 500 kilometres (310 miles) per hour, have even more dramatic beaks to make the most of this effect.

Silent flight

The kingfisher was not the only bird to contribute to the Shinkansen design, though – owls have also had a biomimetic role to play in an unexpected part of the train equipment. Although the tunnel booms provided a significant problem in some locations, the more pervasive noise issue from the Shinkansen tracks was produced by the pantographs – the electrical pickups on the top of the trains that run along the overhead wires to receive electrical energy. These structures make a considerable amount of noise as they generate so-called Karman vortices – a collection of swirls of air travelling in alternate directions behind the pantograph structures as they cut through the air.

The same bird enthusiast who recommended the kingfisher, Eiji Nakatsu, was also aware of the remarkably quiet flight of owls. Where most birds make a fair amount of noise as their wings cut through the air (and some, like pigeons, are extremely noisy), owls are able to approach their prey in near silence, striking without warning. The shape of most birds' wings produces relatively large, noisy vortices from their front edges as they move through the air. But owls' wings have a series of serrations at the front that break up air flow into far smaller vortices, dramatically reducing air noise. By experimenting with pantograph design with serrated edges, the Japanese team were able to significantly reduce the air noise. And yet another bird was invoked – in this case a penguin – in redesigning the shape of the pantograph's strut. This was remodelled to better enable it to cut through the air, similar to the penguin's elegant underwater flight.

If you've ever been in a field when an owl has passed nearby, that near-silent flight that inspired Nakatsu is quite unnerving. A number of years ago, I caught sight of an owl in the headlights of my car on an otherwise very dark night on a country road. The owl was perched on a road sign and didn't seem disturbed when I pulled the car up nearby. I opened the car door and looked down at my phone to get the camera going. When I looked up, the owl had totally disappeared. I was only a few metres away, yet I didn't hear anything as it flew off.

This same ability has been used to inspire an improvement on a more everyday source of noise – not bullet trains, but laptops. As PC processors have got bigger, with billions of connections between components on the chip, the heat they generate has got greater. At one time, this even led to concerns of male user impotence as a result of overheated laps. The generation of this kind of heat almost always means having fans to cool the processors – when a laptop gets under heavy load, the accompanying fan noise can be quite obtrusive. A number of designs for fans have incorporated the front-side serrations used in the Shinkansen pantographs, some with an additional owl-inspired modification.

On the owl's wings, once the airflow has been broken up by those serrations, it travels across the surface of the wing to the back edge. Here, the wings have a more flexible structure than is the case in other birds' wings. This flexibility absorbs energy from the split streams of air, reducing the potential for noise even further. This same approach has been taken to reduce fan blade noise even more.

We have already seen that observations of penguins were added to owl biomimicry in pantograph design. Although penguins are birds, their natural environment is water rather than air – and water may have some different properties to air, but lessons from one fluid can prove beneficial when thinking of another. There have been distinct parallels between the lessons of the owl wing and those learned from the flippers of humpback whales, which are, in effect, the whale's wings for flying through water.

Whales don't stall

Just as the serrations on the owl's wing break up the fluid flow of air, the humpback whale has protrusions on the front of its flippers known as tubercles, which present a lumpy surface to the water flow. In this case, though, there would be a more dramatic impact than reducing noise, thanks to an American aquatic scientist with a truly apposite name – Frank Fish. The effect produced by the tubercles could be used in aircraft design to save lives.

The story starts with a discovery that assumptions can get in the way of creativity. Fish was visiting an art gallery with his wife and saw a sculpture of a humpback whale. The sculpture's flippers had bumps on the leading edges of the flippers. But Fish knew about fluid dynamics, where it was assumed that such leading edges needed to be smooth to avoid creating turbulence. He told the gallery that the tubercles had been put on the wrong side of the flippers, only to be shown photographs proving the sculpture correct. Fascinated, Fish managed to get a 3 metre (10 feet) long sample of a dead whale's flipper, weighing more than 135 kilograms (300 pounds) to take back to the lab to study. He then got his team to put models of the whale flippers into a wind tunnel to see how the tubercles impacted fluid flow. The result was an aerodynamic discovery that shocked experts.

One of the most serious issues facing a pilot is stalling – the sudden dive that occurs when the wings of a plane no longer provide sufficient lift to keep it in the air. There are broadly two causes of stalling – either flying too slowly or angling the wings up at too great an angle in a steep climb. This has a significant impact when a plane is coming into land. It can't safely hit the ground at cruising speeds, so the pilot needs to reduce the velocity through the air to levels that would induce a stall if the wings had been left in the normal flight configuration. To compensate, the wings are extended at the back using extra sections called flaps. These increase the surface area of the wing, one of the principal contributory factors to lift.

What the whale's leading-edge tubercles enable, though, is the ability to achieve a higher angle of climb without stalling. Just as the owl serrations break up turbulent air into smaller, less noisy vortices, Fish's team discovered that whale tubercles reduce the rate at which large-scale turbulence arose as flipper angle increased. This meant that on an aircraft, the wing angle could be roughly doubled before a stall occurred at any particular speed. What's more, stalling is usually a sudden, violent effect – but with wings that are fitted with leading edge tubercles, it is a gradual process.

Although this discovery has potential applications in aircraft that need to suddenly gain height at relatively slow speed – most likely in military applications – in practice, just like the owl serrations, it has found use in fan blades – but on a much larger scale than laptop fans. Adding the equivalent of tubercles to massive ventilation fans leads to a large reduction in energy consumption (around 20 per cent) and in the noise the blades produce. This has also been applied to large fans that are moved by air, rather than employed to push it: the addition of appropriate bumpiness to the leading edges of wind turbine blades makes it possible for wind-powered generators to operate across a wider range of wind speeds, increasing the efficiency of electricity production.

As with owl biomimetics, a second benefit for turbine blade design is noise reduction. At the time of writing, the

UK is considering restarting construction of onshore wind farms. These were suspended for a number of years because of complaints – partly about the visual impact, but also about the noise produced by the turbine blades. Turbine blades inspired by humpback whales, produced by the aptly named WhalePower Corporation, reduce noise as well as being more efficient and able to operate successfully at a greater range of wind speeds.

Honeycombs from structures to atoms

The honeycombs made by bees have provided inspiration for structures for centuries. Using a series of hexagonal cells makes it possible to produce a structure that is both light and strong. Honeycomb structures are used in everything from cardboard to metals and composites, typically with two flat sheets, separated by a honeycomb layer. In the aerospace industry, for example, such structures are often used for the combination of lightness and strength.

More recently, a different natural hexagonal grid structure has led to remarkable new possibilities, operating at the atomic level. One of the natural forms of the element carbon is structured as a series of atom-thick layers of hexagonal grids. This is graphite, something typically used in pencil leads.* It has been known for a long time that these thin layers are quite strong, but only loosely connected to each

---

* The pencil lead has nothing to do with the element lead. It is called this because one of the ores from which lead is extracted is a shiny grey very similar to graphite and the two were sometimes historically confused.

other. This is how a pencil works, as layers of graphite are rubbed off the pencil onto the paper.

However, in 2004, Andre Geim (of gecko tape fame) and Konstantin Novoselov, working at the University of Manchester, were able to separate off individual atom-thick layers from graphite and discovered that this new material, named graphene, which was expected to mimic the honeycomb in its strength and lightness, went far beyond the capabilities of the bees' structure. Initially working by simply separating layers off using sticky tape, the researchers discovered that graphene was ultra-strong and electrically unique.

Because it is just one atom thick, graphene's properties are influenced by quantum effects, leading to many potential applications, from ultra-thin electronics to super-strength materials and specialist filters. While not conventionally biomimetic, the expectation that graphene would be interesting was arguably inspired by its honeycomb-like structure.

Beating the bees

Honeycomb is not the only bee-based source of biomimetic inspiration. As we have already seen with termites, bees are superorganisms where remarkable structures can be built by large numbers of simple, small insects working together as a single unit. This approach has been of increasing interest to the robotics industry.

The most common science-fiction idea of a robot as a mechanical human being has not proved particularly effective for real robotic design. Most industrial robots, for example, are something like a single arm carrying out repetitive tasks, while mobile robots are more likely to travel on wheels than legs, unless specially designed for rough terrain. Biomimetic inspiration suggests, however, that robots might do better if they are based on a swarm of eusocial organisms than a single, complex organism.

Such a design strategy means that individual robots do not have to be as complex as they otherwise would, nor as large. As yet, swarm robots are still very much at the experimental stage, but the potential for them to be useful is significant. One example is in areas where the individual robots have to be too small to have much individual functionality. For example, so-called robots are already used in medical operations, but these are actually the equivalent of drones. They are not autonomous, but simply reproduce their operator's movements on a smaller scale.

These surgical robots are mostly larger than a human. However, if we were to envisage medical robots that operated within a human body, they would have to be tiny, and given the limits of practical miniaturisation, could well need to operate as swarms. Other applications could range from dealing with domestic tasks to military applications, where it has been suggested that an attack by a swarm of small interacting drones, for example, could prove near impossible to resist.

An effective swarm is likely to need some form of artificial intelligence (AI) – and this is reflected in a final structural biomimetic concept of networks.

Catching in a net

At its simplest, a network is a mathematical concept. A network consists of locations known as nodes or vertices and connections between those locations known as edges. (Mathematicians refer to networks as graphs, but 'graph' tends to be used rather differently in ordinary life.) Those nodes and edges can exist in physical space – for example, the structure of the internet and the road network – or virtually. If we envisage, for instance, an airline network, the nodes are real places but the edges are virtual, while other networks can simply be connections between pieces of information.

There are plenty of networks in nature, but arguably the clearest example of an attempt to construct a biomimetic network is in the neural network, a structure commonly used in artificial intelligence. Early attempts at artificial intelligence tried to assemble and structure knowledge, often by attempting to extract knowledge from experts and reproduce their reasoning with a set of rules. This was largely unsuccessful.

More recently, when AI has had successes, it has been through machine-learning technology that makes use of neural networks (we will return to whether the very concept of artificial intelligence is biomimetic in Chapter 7). These were originally based on the way that neurons are linked together and fire (or don't). These cells in the brain are interconnected in a very complex network and function through electrochemical processes at the points the cells connect to each other. Working versions of neural networks, however, diverge considerably from the networks in brains. But this is the whole point of biomimetics – not to slavishly follow nature (as we will see in Chapter 7, this is rarely a good tactic) but to be inspired by it, then develop the approach to fit our particular requirements.

In the case of neural networks, nodes representing the neurons are arranged in layers. Each node is connected to one or more nodes in the next layer, with those connections being capable of being weakened or strengthened as the system learns from data. Values are entered into the first 'input' layer, fed through a number of intermediate 'hidden' layers and finally to an 'output' layer. The network is trained by tweaking the connections to get results that are closer to those required, whether the network is engaged in recognising faces or playing games.

Networks are arguably one of the most abstract structures in which biomimetics has played a part, but the next chapter explores concrete structures that have a very specific role – manipulating light. Here, biomimetics has a number of unusual (and sometimes beautiful) sources of inspiration.

OPTICS 6

Devices to manipulate light, from the simplicity of lenses to the quantum complexities of photonic crystals, often echo the way that light is controlled and modified in nature. Light is, without doubt, the most important energy source in nature. It is light from the Sun that keeps the Earth warm enough for life to exist, that powers the weather systems and that through photosynthesis provides energy directly or indirectly for almost all living things. For that matter, most animals make use of light to see – so it's no surprise that light technology is well-provided with examples of biomimetics – though as always it can be difficult to be sure what is directly inspired by nature.

Take, for instance, solar panels, made up of photovoltaic cells. These convert sunlight into a more practically usable form of energy – electricity. In one sense you could say that a solar panel is a biomimetic development from a plant's use of photosynthesis to gather energy. A plant also uses light energy to move electrons around, though in the biological version this is used in a far more complex process to store away chemical energy. But it is arguable that solar panels are not biomimetic. The ability to generate electrical energy from light using an electronic component was not inspired by photosynthesis – it was discovered by accident. This is more a case of parallel development that happens to have a similar outcome.

For this reason, we won't cover photovoltaics here. But before getting on to more definitively biomimetic inspirations, it's important to consider a more widely used optical application that at least might have some biomimetic stimulation in its origin.

Of spectacles and looking glasses

At its simplest, all optical enhancements, from spectacles to smartphone cameras to telescopes and microscopes are based on biomimetics as the lenses involved may have been initially inspired by nature's lenses in animal eyes. Even the words 'lens' is taken from a shape of nature: *lens* is the Latin for lentil. Early work on lenses dates back at least to the early medieval period (once referred to as the 'Dark Ages', though this term has now been shown to be a huge misunderstanding of a period when plenty of scientific and mathematical advances were made). In the 13th century, for example, the philosopher friar Roger Bacon described numerous experiments using lenses and described what sounds like a very early telescope, though there is no evidence of his constructing such an instrument.

Similarly, although Bacon is often credited with developing the first spectacles that used lenses to correct vision, there is no certainty that this occurred. It is only later that we get detailed evidence of both devices using lenses and of clear biomimetic links to optics. So, for example, Francesco Maurolico, a near-contemporary of Leonardo da Vinci, suggested how a lens focussed light on the retina of the human eye. By the 17th century, French philosopher René Descartes was scraping the retina off a bull's eye to demonstrate this lens work in action.

By the time Descartes had done this, both the telescope and the microscope were in use – though again their origins are not entirely clear. The name often associated with the microscope is Antonie van Leeuwenhoek, who certainly made some impressive observations, including the discovery of some microbial organisms in 1674, but only worked with a single lens instrument – in effect a mounted magnifying glass. To be truly effective, the microscope had to move on to using two lenses. One lens close to the subject of examination produces a magnified image on the opposite side of the lens while the second lens in the eyepiece magnifies that image further. Such compound microscopes are usually attributed to the Dutch father-and-son team of Hans and Zacharias Janssen. Hans started work as early as 1590, though again it is impossible to be sure of an exact date.

Galileo has often been awarded the accolade of inventing the telescope (even by a TV programme as well-researched as *QI*), but we know for certain that his telescope was not the first. Leaving aside Bacon's claims it's entirely possible that Thomas Digges had a crude working telescope in England around the mid-1500s. Not long after, Dutch spectacle makers were getting in on the act.

The name that most often gets mentioned is Hans Lippershey, who attempted to patent the telescope at the start of the 17th century, but was not granted it because others, notably Zacharias Janssen and Jacob Adriaanzoon both had credible prior claims. As for Galileo, it wasn't until he got advance warning of a Dutch telescope being brought to Venice that he scrambled to get his own telescope assembled.

Ironically, for devices intended to help us see more effectively, the origins of these optical aids are anything but clear. We can't even say for certain whether lenses were bio-mimetically based on animal eyes or accidental discoveries using curved glass. It's certainly true that curved pieces of glass and glass globes filled with water were used as burning glasses for starting fires as far back as ancient Greek times, but it is still possible that the understanding of lenses needed to move on to telescopes and microscopes was inspired by a biomimetic exploration of animal eye lenses. Either way, as we'll discover in the next section, having a confused story of where an optical biomimetic concept comes from would continue all the way through to the 20th century.

Origin stories

All the best superheroes have an origin story describing how they came to be something other than an everyday human being. Batman became a grim, costumed vigilante because he witnessed his parents being murdered by thieves (and because his parents had been rich enough to leave him the wealth to fund all those bat-goodies and a playboy life-style). Spider-man was bitten by a radioactive spider, while Superman was dispatched in a probe to cross space from the dying planet of Krypton to Earth, where the different rays of our Sun somehow magically gave him his powers. Unlike

Batman, he had caring adoptive parents to encourage his altogether sunnier disposition.

In their own way, some of the best biomimetic inventions also have origin stories. It's why George de Mestral, who we met in Chapter 1 out hunting with his dog and accidentally being inspired by sticky burrs, makes such a good example. Sometimes, though, there is a suspicion that the origin story is altogether too good. In at least one example of classic optical biomimetic inspiration there is real doubt as to whether the origin story ever happened at all.

This is the tale of Percy Shaw, driving back home from the Dolphin pub along the Queensbury Road, northeast of Halifax in Yorkshire, on a dark night in 1934. The lights of cars back then were not the ultra-bright headlights we now expect, while the road was (and still is) narrow. As Shaw's car rattled along, he came to a turn in the road, which runs along a high ridge with a precipitous drop down towards the town on one side. All that came between Shaw and plunging over the edge was a rickety wooden fence that the car was in danger of crashing straight through as he was unable to clearly see the bend.

Thankfully for Shaw's survival, there was a cat perched on the fence, staring down disdainfully at the approaching car. The retinas of some animals' eyes are particularly good at reflecting light. This is because they have a layer behind the retina called the tapetum lucidum that reflects light back, so the retina's sensors have a double chance of detecting low-level light. This layer tends to be found in animals that are nocturnal predators. The cat's eyes glowed green in Shaw's headlight beams against the dark backdrop. He realised, just in time, that he needed to turn to avoid plunging to his death.

Inspired by the cat's protective presence, Shaw spent many months devising a road marker that would have the same effect, able to reflect a car's headlights back to the car, showing up the edge or middle of the road. Such a marker would require no power, making it cheap and meaning that it would not need complex infrastructure to keep it lit. At the edge of the road, like the cat, a reflector could be at a higher level, but to mark the centre of the road, there was an additional problem. How to prevent the reflector from becoming covered in mud.

A number of years later, Shaw appears to have had another biomimetic inspiration. In nature, many animals keep their eyeballs clear by blinking – moving an eyelid over the moistened surface to carry away debris. In Shaw's devices, which are still to this day called cats' eyes in the UK, a kind of artificial rubber eyelid automatically wipes the reflective lenses. They use moisture from rainwater that is collected in a little reservoir, wiping the lenses clean when a car drives over them. The products of Shaw's new company, Reflecting Roadstuds Limited, would prove hugely popular on unlit roads, given a boost by the dark streets of the Second World War when blackouts were used to help defend against bomber raids.

It's a wonderful story. And it is entirely possible that Shaw did have some inspiration from nature. Anyone who has seen an animal's eyes glowing eerily by reflected light in the dark (oddly, it's rarely a cat because they are usually too low to reflect well) might conceive of such an approach. But the detail of the life-saving cat on Shaw's perilous journey is very likely to have been a confection. He told several widely differing versions of the inspiration, sometimes suggesting the reflector that got him thinking was a metal tram track and at others that there were already eye-level reflectors on road signs but driving in fog made him realise how valuable they would be if they were embedded in the road.

It's hard to be certain – but Shaw's cat's eyes likely had at least some biomimetic inspiration. And that is certainly true with a concept known as structural colouration. First, though, we need to be clear what this involves.

Colour from structure

The previous chapter focused on structures – but there is one kind of natural structure that has the role of producing an unexpected optical effect, known as structural colouration. This is when a substance can appear to have a specific hue not because it contains pigments that produce that colouration, but because the structure of the object affects light in such a way to generate particular colours.

The basics of light and the usual way that an object appeared coloured was worked out by Isaac Newton in his letter 'On Colour and Light', written to Henry Oldenburg, who was then the secretary of the Royal Society of London. It is here that Newton describes his work with prisms that would identify the nature of white light as being a combination of the colours of the rainbow, which are split out when the light passes through a prism.

Newton goes on to explain how an object appears to have a particular colour. He describes two pigments, minium (a lead oxide), which was used to produce bright reds in the Middle Ages, and bise, a blue pigment based on copper carbonate. In each case, the object was visible when he lit it with any colour of light but was 'most luminous' in red and blue light respectively. He concluded that the colour we associate with an object or pigment is the shade of the light that it most strongly reflects, while it absorbs more of the other colours present in white light.

This letter dates from 1672 – and in it, Newton refers to a book called *Micrographia* by his arch-rival Robert Hooke. A significant scientist in his own right, Hooke, who was curator of experiments at the Royal Society, attacked Newton's letter on a number of fronts. Newton turned out to be right, but in *Micrographia*, written in 1665 before Newton's explanation of normal colouration, Hooke had already put forward a suggestion of how structural colouration could work.

Micrographia is now primarily remembered for its outstanding, huge illustrations of the likes of a flea, a louse and the compound eyes of a fly as seen through a microscope. But in a section titled 'Of the Colours observable in Muscovy Glass, and other thin Bodies', Hooke begins by describing how the mineral muscovite (then called Muscovy glass as thin layers of it were used in Russian windows), a substance that is of itself pretty much colourless, could exhibit a range of colours. Hooke explains how he saw 'Blew, Purple, Scarlet, Yellow, Green; Blew, Purple, Scarlet, and so onwards, sometimes half a score times repeated, that is, there appeared six, seven, eight, nine or ten several coloured rings or lines, each incircling the other'.

Hooke goes on to express the hope that one day it will be possible to find the causes of such phenomena and whether studying them would make it possible to 'deduce the true causes of the production of all kinds of Colours'. This led him to also mention the colours of peacock and other feathers. By experimenting with these feathers, Hooke was able to suggest that their colours came from refraction and reflection effects. This was because when he wet the feathers this 'destroy'd their colours'. What had been magnificently colourful were now 'ting'd with a darkish red colour, nothing akin to the curious and lovely greens and blues they exhibited'.

Newton took umbrage at the attack on his letter and would not publish his complete ideas on light until his book *Opticks* came out in 1704. Here, he went further than Hooke in ascribing the source of the colours of the peacock's tail to the thinness of the transparent parts of the feathers. Later observers, such as the English polymath Thomas Young, would realise that what was happening in such examples of colour production was interference.

If we think of light as a wave (an idea that Young was responsible for making the definitive model of light until Einstein's day), interference results from two or more waves of light interacting. If two superimposed waves are both rising at the same time and location, they will reinforce each other, producing a stronger wave. If, on the other hand, one is rising and one is falling, they will interfere destructively, cancelling each other out.

Thin films, like muscovite crystals or the structures in peacock feathers, enable the light reflected from the top surface of the film and the light reflected from the bottom surface to interfere with each other. The thickness of the film will determine which colours are reinforced and which are cancelled out, producing colours that are not dependent on a pigment. Familiar examples of such thin film interference are the colours seen in a soap bubble and in a thin layer of oil on the road. Such colouration, dependent on the angle of view, is often known as iridescence.

Some butterflies, such as the intensely blue *Morpho didius*, get around the directional limitations of iridescence by having complex, fern-like structures on their wing scales. These structures produce the colour in many different directions because of their irregular shapes, so have structural colouration that is not considered iridescent.

Another mechanism of structural colouration that occurs in nature is the photonic crystal, made up of a series of tiny structures of a similar scale to the wavelength of a particular colour of light. Such structures tend to strongly reflect colours of around that wavelength, producing the appearance of an intense colouration. For example, some butterflies have arrays of tiny holes in their wings which can produce an intense reflection, drastically change the angle at which different colours are refracted* or block certain colours entirely.

Is this the real colour? Is this just fantasy?

It's tempting, incidentally, to think of there being 'real' colours that objects have somehow inherently produced by their pigmentation, and 'fake' colours that are the result of structures. The author Thomas Mann provides us with a character in his novel *Doctor Faustus* who thinks this, commenting that the azure colour of some organisms was 'not a real or genuine colour, but was produced by delicate grooves and other variations'. However (as another Mann character explains), this totally misunderstands the nature of being coloured.

---

* Refraction is the bending of light as it passes from one medium to another – for example, when it passes from air to water. Different colours refract at different angles: this is how rainbows are produced.

There are two different ways to look at colouration, the existence of which causes confusion to philosophers, some of whom take what is arguably the wrong viewpoint. If you imagine something that you describe as being red, you could either be thinking of the sensation produced in your brain that you call seeing something red, or you can imagine the production of light photons in a certain band of energies (or wavelengths if you prefer).

If we take the 'sensation produced in your brain that you call seeing something red', we get into all sorts of philosophical difficulties. We have no way of knowing if someone else sees the same thing when they describe something as being red. I can only know what I experience when I see something I think is red. You can only know your own experience. We assume those experiences are similar – and for most people who aren't colourblind, that may well be the case – but we can't prove that the experiences are identical.

However, I would argue that the true colour of an object should be simply defined by the energies of the photons it gives off. Assuming that is the case, it really doesn't matter how those photons are produced. They could be the photon energies that are re-emitted from white light hitting a pigment. But those photons could equally be produced by interaction with a structure. Or, for that matter, as a result of the scattering process that makes the sky blue. It makes no difference how the photons are selected or produced – the result is the colour of the object we see.

It doesn't help in trying to untangle this that we generally consider the colours of objects to change under certain circumstances, but not under others, even if the colour appears different as far as our eyes are concerned. We don't say that the sky is blue at night: we consider it has changed to be black, even though the sky itself is physically unchanged. Contrast this with what happens when lighting conditions get particularly dim. The cones in our eyes that detect colours stop working, leaving only a monochrome view from the more sensitive rod cells. Despite this, most of us would say that, for example, a red car is still red, even if it appears to be grey in low light. Arguably, structural colours tend to be treated more like the sky than the car. If, for example, a bird's blue feathers look black when they are wet, we would tend to say that they have undergone a colour change in a considerably different way to the change in appearance of that red car.

Manipulating photons

Photonic crystals and the related technology known as metamaterials have certainly been employed in biomimetic applications. The idea of a glass lens as the way to focus light is so firmly fixed in our minds that it can be easy to forget that all focusing involves is a change of the direction in which photons are travelling – and glass is not the only material that can do this. With our understanding of the quantum nature of light, and the ability to manipulate materials on an extremely small scale, optical devices that far exceed the capabilities of glass lenses can now be made, using materials that we would not normally think of as transparent.

The two principal forms of quantum lens are meta-materials and photonic crystals. Each modifies the way light behaves by interacting with individual photons. Metamaterials have special properties that are produced by the way that they are assembled. The optical effect is produced by a complex structure, which might be a pattern of tiny holes or layers of lattices that collectively interact with photons. It is this structure, rather than chemical composition as with pigments, that gives them their properties. Photonic crystals can be arranged to act on light in the same way as semiconductors act on electricity, giving unparalleled precision of control.

Although, as we have seen, some butterfly wings act as metamaterials, the artificial versions go beyond anything found in nature. We would normally expect a transparent material to have a positive refractive index. This means that when light hits something transparent, like a block of glass or a mass of water, it bends inwards, towards an imaginary line going straight into the material. However, when light hits a metamaterial, it is bent in the opposite direction, away from the line. The metamaterial has a negative refractive index. This might seem a trivial difference, but it means that metamaterials can manipulate light in ways that feel unnatural to us.

One practical application is the ability to go beyond the limit that is imposed on normal lenses by the wavelength of light. There is a scale below which no conventional optical microscope can focus, however powerful it may be. If you try and observe an object that is smaller than the wavelength of the light used to view it, the result is inevitably failure. The wave effectively bypasses the object rather than reflecting off it. But this limitation is removed when super-lenses made from metamaterials are used. These can take optical focus down to detail that was only previously detectable with electron microscopes. Not only could such metamaterial lenses be built for a fraction of the cost of an electron microscope, they enable a different kind of observation, just as radio telescopes and visible light telescopes can work together in astronomy to get a more complete picture.

As we have seen, unlike the more sophisticated types of metamaterials, photonic lattices do occur in nature, though only in imperfect forms that can't be used as lenses. Both the swirly, glittering appearance of an opal and the iridescence of a peacock's tail are caused by photonic lattices. But the practical applications of artificially created photonic crystals can do much more than produce a pretty effect. Because a photonic lattice acts on light as semiconductors do on electrons, they are essential components if we are ever to build optical computers.

These theoretical machines would use photons to represent bits, rather than the electrical impulses used in a conventional computer. This could vastly increase computing power. Because photons don't interact with each other, many more can be crammed into a tiny space, or can flow across each other in a basket of light, allowing more complex and faster architectures. Equally, optical switching – and in the end, a computer is just a huge array of switches – could be much faster than the electronic equivalent. There are significant technical problems to be overcome, but the potential is great.

Colour me photonic

There are no super lenses in nature. However, a direct biomimetic inspiration has been to look at the alternatives that photonic materials present to pigments for colouring materials. Teijin, a Japanese company, produced a fabric in 2007 based on the Morpho butterfly's photonic colouring

unimaginatively called Morphotex. According to the manufacturer, 'Thin films of 70 nm thickness consisting either of
polyester or nylon are laminated in 61 layers alternatively,
and four types of basic colors such as red, green, blue and
violet are allowed to be developed by precisely controlling
the layer thickness according to visible wavelength.'

However, despite use in the seat fabric of a short-lived
special edition Nissan convertible and some success at Japan
Fashion Week for a dress made in the fabric by designer
Donna Sgro, the material did not make the mainstream.
According to the designer, the fabric eliminated 'the highly
toxic process of industrial fabric dying ... my garments
demonstrate the possibilities for engaging technological
solutions to the problem of textile waste, encourage dialogue
around the issue of sustainable fashion, and links biomimicry, an emergent practice within the field of sustainability, to
fashion design.' While it's true that there are environmental
benefits available from not using dyes, it's not entirely clear
how the dress does anything for the problem of textile waste
(or what biomimetics have to do with sustainability). Either
way, the fabric does not seem to have been a commercial
success: Morphotex does not appear on Teijin's website.

Similarly, cosmetics company L'Oréal has experimented
with using photonic crystals in its products. Back in 2005,
the company announced that it was working with Professor
Pete Vukusic at Exeter University on incorporating structural colouration into its products. According to the L'Oréal
research director Patricia Pineau, 'This is amazing when
you consider the basic colour of the product is just white.
It brings great advantages for cosmetic applications, being
particularly beneficial in lipsticks, as it means that any transfer of the product onto another surface only leaves a white

powder trace, not the colour.' It's not entirely clear why the structural colouration would work on lips, but not on clothing.

As yet, as seems so often the case with a biomimetic product, the use of structural colouration to replace pigments is very much the equivalent of a concept car for the fashion and cosmetics industries – these are interesting ideas, but they have not translated into everyday products. When I asked L'Oréal about actual products on sale incorporating this technology they were not forthcoming – and there is no mention of photonics on their website. The idea is eye-catching, but the practicality is lacking. (It's also possibly the case that the colours produced were too garish for much of the industry.)

The visible invisible

In nature, all is not what it seems visually – at least, not how it appears to our eyes. Other animals are able to see beyond the part of the light spectrum we can pick up. This can be both the lower energy infrared and the higher energy ultraviolet. For example, many flowers from sunflowers to geraniums have patterns on their petals invisible to us that help guide insects in to pollinate them. Similarly, some birds of prey have a fourth type of colour-detecting cone in their eyes. The human eye has three different cone types which, between them, cover the range of electromagnetic radiation we think of as visible light. But to these birds, light we simply can't see is as distinct as any visible colour to us.

When, for example, a hawk swoops down to catch a mouse or shrew, it is quite possible it can't see the animal itself, camouflaged against the grass. Instead, it is picking up the trail of urine these small mammals leave behind, which glows in the ultraviolet light from the Sun. To us there would be nothing to see – to the bird of prey, there is a glowing pointer to its prey (provided it guesses correctly which way the animal is running).

It might seem that a biomimetic application of something we can't see makes little sense, but that's exactly what happened in 2012 when the lookout tower on Lindisfarne was renovated. Also known as Holy Island, Lindisfarne is an island off the coast of Northumberland in the north-east of England, accessible by causeway at low tide. It is home to the ruined Lindisfarne Priory and a castle. On the south coast, near the priory, is a lookout tower, giving views across the island and out to sea. It was originally built as a watchtower for coastguards to monitor boats in storms, but now has been turned into a tourist attraction.

When the tower was rebuilt in 2012, converting it into an elegant modern structure, it was given a glass lantern on top to provide clear views all around. But this part of the coast is home to important bird sanctuaries, and it was a concern that the large glass windows would prove to be a flight hazard. Birds usually struggle to see window glass, and because the lantern meant that there would be a view straight through the building, it seemed likely that many birds would fly straight into the glass, potentially killing them (and also making a bit of a mess).

The architects were the first in the UK to make use of a German biomimetic product – ORNILUX glass, made by the company Arnold Glas. To the human eye, the glass looks perfectly normal, unless it is closely inspected. On the surface is a coating that looks, when observed with ultraviolet light, like an array of criss-crossing bamboo rods. This is referred to as a 'mikado coating', after the German name for the game of pick-up sticks.

The biomimetic basis here is the web of the orb spider. (The pattern has been described as looking like a spider web, but this confuses the inspiration with the actual design used.) The webs of the orb spider have extra, ultraviolet reflective strands known as stabilimenta. The fine webs are hard for flying birds to spot, but these glowing patterns make the webs stand out and prevent birds from flying through them and destroying them. Tests of the new glass at the Max Planck Institute for Ornithology showed that over 60 per cent of birds that would otherwise have flown straight into a conventional sheet of glass avoided the treated product.

In principle, this is dealing with a big problem (for birds, at least). It has been suggested that around 1 billion birds die each year from collisions with buildings, primarily where there are large sheets of glass.* This is one of those statistics that seems to get everywhere without there being good evidence to back it up. The 1 billion figure was probably based on an estimate for the US of between 100 million and 1 billion birds. These numbers came from a rough calculation made by biologist Daniel Klem in 1990. He estimated the number of buildings in the US at around 97 million and that between one and ten birds a year would be killed by each building. A broad guesstimate at best.

There is no doubt that wider use of this type of glass would reduce bird collisions, though in what is now a familiar

---

* This sounds a huge number, but even more are killed by domestic cats, which are estimated to kill between 1.3 and 4 billion birds, plus between 6.3 and 22.3 billion small mammals each year.

story with biomimetics, at the time of writing it has not been widely implemented as the extra cost is rarely seen as being justified (and many don't even know the product exists).

Hiding in plain sight

One potential application of metamaterials is making objects invisible. Because of their negative refractive index, metamaterials can bend light around an object, acting as an invisibility cloak. This has already been done on a very small scale with microwaves, but is harder to make practical with visible light as the materials used absorb too much of the light to work effectively. As with metamaterial lenses, this isn't biomimetic as there is no natural equivalent, but there is in a more subtle way of making things disappear where we have learned a lot from nature: camouflage and mimicry.

Camouflage is a surprisingly modern term, taken from the French word for disguise and only in common use since the First World War. We can learn a lot from natural examples when it comes to developing camouflage – not only in discovering how different camouflage mechanisms work and using them on our equipment and individuals that we want to conceal, but also in the ways that animals use camouflage not to disappear, but to look like something that they aren't. This could be anything from a stick insect appearing to be part of a bush to comma butterflies, which have wings that provide a near-perfect image of a dead leaf, down to a white splodge of bird droppings and edges that look torn.

In other examples, animals will do the opposite of these attempts to disappear in the background and will instead deliberately stand out to emphasise the fact that they are dangerous.

Think, for example, of the bright yellow and black livery of a wasp, that broadcasts 'avoid me or you'll get stung'. This then results in one of the aspects of evolution that fascinated Charles Darwin – mimicry. Harmless hover flies come in species variants that look like most types of wasp or bee (including fake bumble bees). They have no sting, but insects that look more like an aggressor are less likely to be eaten, so more likely to survive and pass on this appearance to their offspring.

From human wasps to human stick insects

Something quite remarkable happened to the military towards the end of the 19th century. Up until this point, soldiers had been outfitted in a way that took the same approach to appearance as wasps. They aggressively advertised that they were present and that they were dangerous. Soldiers' uniforms were bright colours – often red, arguably (alongside yellow) the most visible colour against the greens and blues of natural backgrounds.

This probably made a degree of sense when the weapons of choice were short range – swords, for example. However, as guns became more prevalent and more accurate at long distances, the designers of combat wear took a different lesson from nature. A better survival strategy is not to be seen, but to blend into the background, able to then shoot an enemy that becomes visible without being spotted yourself.

The disappearing ships

The first evidence of intentional biomimetics in military camouflage arose from the American–Spanish war of 1898.

The US Navy asked American artist Abbott Thayer to give them guidance on making their ships less visible. Thayer had been interested in the way that in nature some animals used countershading to be less visible. The theory behind this was derived from the way that sunlight naturally comes from above. If an animal is the same colour all over, the more brightly illuminated top part would be brighter than the bottom part. As a result, Thayer and others suggested, many animals are darker on the top than they are on the bottom. (Think, for example, of the shading on a shark or a deer with their lighter undersides.) This 'countershading' is thought to make the animal less clearly visible in natural light.

Despite being called in by the navy, Thayer's ideas of making naval vessels darker on top, shading through to white beneath, were not taken up. It would instead be the First World War that saw the first practical deployment of biomimetic camouflage, though this was more based on a second type of animal camouflage that Thayer also championed, known as disruptive colouration,* which takes a different approach to the imitative colour scheme often employed by insects.

It is relatively easy for an insect to blend into, say, tree bark as it is much smaller than a tree and because it typically spends a lot of time not moving. Larger animals tend not to be able to blend in so effectively – you don't see big animals with bark-like colouration, for example. Instead, they are more likely to display patterns such as stripes, patches or spots that break up their outline. This doesn't render them invisible but makes it less likely that the animal will be identified as what it truly is.

---

* Thayer actually called it ruptive colouration.

There is no doubt that disruptive camouflage really exists in nature, though Thayer was so obsessed with the concept that he took it to ridiculous lengths, suggesting that, for example, the bright patterns on skunks or wasps were there to make them difficult to be identified, rather than to warn predators off as is otherwise universally acknowledged. (Interestingly, Thayer's greatest critic on this subject was US President Theodore Roosevelt, who after his period in office seems to have delighted in pointing out flaws in Thayer's logic.) Disruptive colouration is a real thing but does not apply to every pattern on an animal – and in some cases, such as a zebra,* the apparently disruptive colouration doesn't seem to work very well.

Disruptive colouration came into play on ships during the First World War with the introduction of bizarre looking patterns. They certainly did not hide the ships – quite the reverse.

The extreme disruptive colouration used, known as dazzle camouflage (Thayer called it razzle dazzle) broke up the outline of the vessel with stark black and white stripes to such an extent that it was almost impossible to work out the direction the ship was facing in. This made it less susceptible to attacks by submarine torpedoes, which have to anticipate where a ship will be some time after firing. Using camouflage on ships was championed in the UK by the zoologist John Kerr, who sent Winston Churchill (then First Lord of the Admiralty) a briefing on the approach in September 1914. Kerr recommended both countershading

---

* The zebra has been subject to many theories as to why it has its stripes. None is particularly satisfactory – for the moment, the simple answer is we don't know.

The USS West Mahomet in disruptive dazzle camouflage.
Naval History & Heritage Command, Public domain, via Wikimedia Commons

and animal-style disruptive colouration. However, the more dramatic (and less scientifically based) approach of dazzle camouflage came from the artist Norman Wilkinson.

Whether this approach was effective remains much disputed. It was pointed out that the dazzle camouflage made the ships easier to spot, particularly on a moonlit night, and made them arguably more susceptible to attacks from guns, as opposed to torpedoes. By the end of the war, over 2,300 merchant vessels had been painted in dazzle camouflage. The statistics on sinkings were inconclusive – over some periods a higher percentage of camouflaged ships were sunk, on others more of the conventionally painted vessels. The benefits were not clear – but the principle of camouflaging ships had moved from theory to practice.

By the time the Second World War came around, there was more enthusiasm for camouflage of ships. Led by the naturalist Peter Scott, there was a move for British warships to adopt a mild disruptive camouflage in patterns using blues, greens and whites. By 1941, some scientific research had been done, showing that in dull lighting, white or very pale colours were far harder to spot than the darker greys or black then typically used for ships. Nothing was perfect in all conditions and locations, but light colours with some disruptive patterns seemed to have the best general result. At the same time, the US Navy opted for more dazzle-oriented camouflage for its ships, designed to confuse range and direction of travel.

Spot the soldier

Naturally inspired camouflage was not, of course, limited to ships. Soldiers on the ground – and their equipment – would clearly benefit from concealment. Unlike the bright red uniforms of old, a move was made towards more natural colours (as had been used for centuries by countrymen wanting to limit detection by game). Thayer was involved in this area too, suggesting khaki with both disruptive colouration and scraps of cloth to break up outlines. This idea was not accepted everywhere: even basic camouflage patterns were considered to go too far for the mass of infantry, though the benefit was seen for snipers. From 1916, British snipers had suits that used strips of cloth in camouflage, though these seem to have been based on the outfits of Scottish ghillies,* rather than Thayer's designs.

Interestingly, although there was no doubt that camouflage was inspired by nature, and therefore genuinely biomimetic, many of those responsible for First World War camouflage were artists rather than naturalists. This meant that much early camouflage seemed more inspired by the art trends of the time, notably cubism – and were far more brightly coloured than would be the case once there was more of a scientific input to the design.

At the start of the Second World War, there was little in the way of camouflage for ordinary soldiers on the ground, but over time more combat gear was introduced that used leaf-like patterns in an attempt to reduce visibility. As both the Germans and the Allies did this, there was a certain amount of confusion caused when it became difficult to distinguish which army a soldier was fighting for.

Into the air

By the start of the Second World War in 1939, the impact of aerial warfare was an essential consideration. This brought new requirements of camouflage – both in trying to conceal

---

* Professional deer stalkers.

planes in the air and trying to hide airfields and other critical targets from enemy aircraft.

There was never any successful conclusion for what would work for the planes themselves. Their high-speed motion and the need to conceal them against a range of back-drops depending on the angle they were seen from resulted in a mess of different partial solutions. British planes had brown and green patches on top (almost useless in motion) and pale blue undersides to match the sky. German aircraft also had pale blue undersides but were mottled blue on top. And the US simply gave up on the idea, leaving their planes unpainted silver.

There was more success with airfields, though devel-opment of camouflage here in the UK was disrupted by disputes between the zoologist Hugh Cott (who had worked with John Kerr) and a Colonel Turner, who had Norman Wilkinson in his employ. The general approach taken by both sides was netting raised above structures carrying painted and natural camouflage material, but there was considerable dispute over detail. Despite success in a small-scale trial, Cott was initially sidelined.

One of Turner's ventures was the production of fake air-fields – reproducing the natural approach of mimicry in a rather different way. Dummy planes were mocked up by film studios for daytime fakes, while lighting was used at night to suggest the presence of imaginary airfields. These were not trivial contributors to the war effort. During the Battle of Britain in 1940, twice as many attacks were made by the Germans on fake airfields as on real ones.

By 1940, Cott was ready to fight back, having produced a definitive book on natural camouflage, *Adaptive Coloration in Animals*. He pointed out the lack of effectiveness of much of the camouflage of the time, often consisting only of patches of shiny materials in earth colours. These were deployed even where they made no sense – for example, on the tops of London buses. Not only was the camouflage unsuited to an urban situation, the buses retained bright red paint on their sides. Cott pointed out that too much consideration was being given to the views of artists and not enough to those who could successfully mimic the work of nature.

Cott's views were published in the journal *Nature* and soon after he was allowed to make a comparative camouflage effort on a pair of rail-mounted guns. One was given the usual earth colour lozenge pattern. The other made use of biomimetic countershading. Aerial photographs were taken, and Cott's countershading worked far better, rendering the gun practically invisible, while the traditional camouflage was useless.

The turnaround was not immediate, but Cott was now included with painters (and the magician Jasper Maskelyne) as a contributor to a military camouflage development and training centre. Biomimetics was at the heart of the training given by Cott, stressing that unlike humans, animals in the wild are highly dependent for survival on blending into the environment.

Disguising in the desert

One of the biggest challenges faced by Cott and his colleagues was the move to North Africa, where most existing camouflage materials had to be reconsidered, as what worked in a European field was not suitable for a desert. One explicit example of a lesson from nature was to make use of a technique found on some chrysalises or insect wings of appearing to already be damaged beyond being interesting to predators. A water distillery at Tobruk was given the appearance of already being blown apart to deter future raids.

One of Cott's colleagues even suggest an option inspired by the wing spots on some butterflies. These spots look like eyes – they encourage a predator to aim for a less fatal part of the body. The idea was never tried in practice, but the suggestion was to disguise the tenders of steam locomotives as boilers. Air raids tended to target loco boilers as the most fragile part of the locomotive – this might have resulted in fewer trains being put out of action.

This was the first idea in a direction that would be carried further towards those insects like hover flies and stick insects that give the impression of being something entirely different to their actual nature. This biomimetic approach would be brought into use for the Battle of El Alamein in 1942, when guns and tanks were very successfully disguised using canvas superstructures to look like trucks. Part of the effectiveness of this approach was being aware of the key differences between, say, a tank and a truck. A tank would leave distinctive tracks in the sand, so the camouflage rig included a piece of equipment dragged behind the tank that wiped away its tracks.

This biomimetic approach would be undertaken with much higher levels of technology in later years. Defence tech company BAE Systems developed a system known as ADAPTIV which deals with the problem of a tank's easy visibility to infrared night vision goggles. The metal hull of the tank is sufficiently hotter than its surroundings that it stands out like a sore thumb. Hiding the thing entirely is nigh-on impossible, especially when it moves. Accordingly, the ADAPTIV system changes the temperature of hexagonal panels on the side of the tank, which are used to transform its appearance. Through night vision goggles, what is seen is not a tank but a family car – even featuring wheels and windows.

A similar deception approach is undertaken with stealth aircraft. It's pretty much impossible to make an aircraft entirely disappear to light or radar. The stealth technology reduces reflection, but it is also designed to change the apparent shape of the aircraft so that it seems much smaller than it actually is.

Disappearing building

One last example of experimental optical biomimetics comes not from the military but from cities – and has been inspired by flatfish. We tend to think of chameleons as the natural champions of disguise, but in reality, most species of chameleon don't use their changes of colour to hide at all – it's all about signalling to other chameleons. However, there are colour-changing fish like soles that truly can blend into the background.

The fish combine the colour-changing cells known as chromatophores used by chameleons with sensors on their undersides that enable them to pass on a rough approximation of the pattern of the seabed to their upper surface. This approach has been proposed to be used on a South Korean tower block called Tower Infinity.

The idea was that the tower, to be built near Seoul airport, would be covered in video cameras that would feed the image they were picking up to LED screens on the opposite side of the tower. When looking at the tower, what you were able to see was the view from the other side of the tower – the building would vanish (or at least become distinctly fuzzy). This design was accepted by the government in 2013. Construction was started at the end of 2019, but at the time of writing the tower is yet to be completed ... so we don't know how well this flatfish-inspired system works.[*]

Throughout this book so far, we have seen good examples of biomimetic ideas, though with the proviso that they are rarely large-scale successes, often being either one-offs or of limited use in the real world. It would be useful then to consider just what the limitations of biomimetics are and whether we sometimes fool ourselves about whether a product really is imitating life or the natural world.

---

[*] Whether or not it's a good idea to have an invisible tower near an airport is a different issue that appears not to have been discussed.

THE COPYCAT TRAP 7

Biomimetics is a beguiling approach to product development, tying into the 'natural is good' appeal so often used in product marketing, but as we have seen, copying nature always needs to come with a health warning: natural is not always best. Evolution enables nature to develop solutions to problems, and it has had a far longer timescale to work on things than has human ingenuity – but intelligence can be far more efficient than evolution because our thought processes are goal driven. When it comes to biomimetics, then, we should not be looking to copy what nature does, but to learn from nature and combine natural inspiration with human intelligence and design. As the American philosopher of technology Lewis Mumford has commented: 'The most ineffective kind of machine is the *realistic* mechanical imitation of a man or another animal.'

Taking to the air

Although it seemed the obvious thing to do, and was attempted by early inventors, imitating the flapping wings of birds does not make for a useful flying machine. In one sense, almost all aircraft are biomimetic. Most flying machines do *have* wings (and those that don't use wings conventionally, such as helicopters, still incorporate some of the basic principles of flight found in nature, such as the rotating wings of a falling sycamore seed). There is definite biomimetic inspiration for the use of wings in gliding, as anyone who has watched a soaring red kite glide for minutes at a time without a single movement of its outstretched wings will realise.

However, most early attempts at flying machines did assume that the natural way forward for powered flight was to have a human or machine flap wings like a bird. A familiar tale from ancient Greece features the escapades of Daedalus and Icarus. Daedalus was the mythical brains behind the Cretan labyrinth of King Minos. According to the story, the inventor told Minos' daughter Ariadne how to get through the labyrinth using a ball of thread. She handed the solution over to her suitor, Theseus, who promptly ventured into the maze and slaughtered the labyrinth's resident bull-man, the Minotaur.

This didn't go too well for Daedalus, who was put in jail along with his son, Icarus. The great inventor was able to produce two pairs of wings made with wax and feathers (this clearly wasn't a maximum-security jail) – which enabled the pair to fly away until Icarus ventured too close to the Sun, melting the wax and leaving him to plunge to his death in the sea. This was all intended as an instructive myth on the topic of hubris, rather than a lesson in how to construct flying machines, but it seemed reasonable for many centuries that such devices should be constructed making use of flapping wings.

One of the earliest recorded attempts to replicate the flying feat of Daedalus was made by a monk from Malmesbury Abbey in Wiltshire. Eilmer (sometimes known as Oliver) made himself wooden wings with feathers attached and, with remarkable courage of conviction, jumped off the top of the Abbey tower in 1020. The tower was topped with a spire taking it up to 131 metres (431 feet), though the accessible summit was probably no more than two-thirds of that height. Although Eilmer inevitably failed to soar like a bird, it seems likely that his wings provided some uplift. Admittedly he broke both his legs, but he did survive.

By the mid-13th century, the medieval proto-scientist Roger Bacon we met in the optics section would write in his *Letter Concerning the Marvellous Power of Art and Nature and Concerning the Nullity of Magic*: 'It is possible that a device for flying shall be made such that a man sitting in the middle of it and turning a crank shall cause artificial wings to beat the air after the manner of a bird's flight.' This was included in a list of remarkable devices, though it did come with the following qualifier: 'These devices have been made in antiquity and in our own time, and they are certain. I am acquainted with them explicitly, except with the instrument for flying, which I have not seen. And I know no one who has seen it. But I know a wise man who has thought out the artifice.'

Bacon's honesty that he hadn't seen the flying machine was justified, because it was one thing to 'think out the artifice' and another to make it practical. Birds are able to fly by flapping their wings thanks to a combination of having very light, hollow bones and disproportionately large muscles. Human body weight is simply far too great for our musculature to ever be able to generate enough uplift by flapping – even though 'bird men' would attempt human-powered flight that wasn't dependent on gliding for lift all the way through to the end of the 19th century (and occasionally still do today, though the approach is probably now largely ironic).

There was another attempt made at biomimetic flying machine design a little over two centuries after Roger Bacon from Leonardo da Vinci. As far as we know, Leonardo never attempted to build aircraft, but he certainly made a good attempt at designing them. Although his first area of expertise was art, Leonardo bluffed his way into the employment of the Duke of Milan by claiming to be a skilled military engineer. Despite a lack of previous experience, his mechanical designs were fascinating. Several of his sketches suggest something like Roger Bacon's hand-cranked flapping device, but more interestingly, he played with the idea of experimentally testing how much lift could be gained when gliding from a Batman-like scalloped fixed wing.

Most famously (and particularly relevant here), Leonardo came up with a design for what has, rather optimistically, been described as a helicopter. Rather than using wings, this attempted to fly by using a large helical screw to cut through the air – it's entirely possible this was inspired by the wind-borne seeds of trees, such as the sycamore, which have aerofoils that produce a spinning action. All told, Leonardo scattered his notebooks with 150 sketches of flying machines – something that clearly appealed strongly to him.

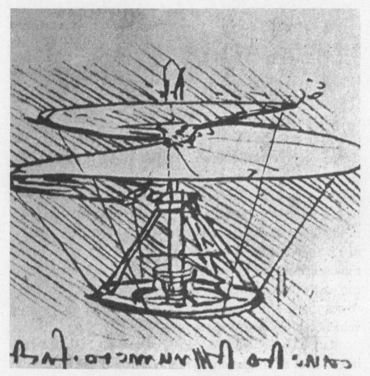

**Leonardo's 'helicopter' design seems
influenced by wind-borne seeds.**
Leonardo da Vinci, Public domain, via Wikimedia Commons

Gliding to success

The most significant biomimetic success for flight, which surely can be attributed to watching birds soaring with wings fixed wide, was in the development of gliders. The English inventor Sir George Cayley first advocated fixed-winged flight in the early 1800s, but, anecdotal evidence apart, it wouldn't be until the 1890s that German inventor Otto Lilienthal took to the air, gliding with a pair of wings that initially bore a resemblance to birds' in the shape, though later became more stylised. Lilienthal had considerable success until 1896 when he was killed as a result of losing control at 15 metres (50 feet) above the ground.

His successors, who in 1903 would become the pioneers of powered flight, the Wright brothers, were aware of the dangers of Lilienthal's approach – particularly in terms of the way that he controlled the direction of flight (this was probably the cause of his fatal crash). Wilbur Wright, in particular, had an interest in birds: biomimetics would come into play in his inspiration for a different way to direct the motion of a flying machine through the air. It's hard to get into the mindset of the time – we are so used to the way that planes can easily change direction in mid-air that it can be difficult to see that it's not at all obvious how to make an airborne vehicle change course.

Rather as a skateboarder does, Lilienthal had changed the direction of flight by shifting his body from side-to-side. But Wilbur had noticed that birds made use of a more sophisticated directional control by changing the shape of their wings, twisting the outer tips to point in different directions. To mimic this, the Wright Brothers rigged cables that enabled them to curve the wings of their planes – 'wing-warping' as it was called. After experimenting with kites, they added extra features that would become known as control surfaces – a rudder to compensate for the way that wing-end twisting tended to change the direction the plane faced in and an elevator at the front (later moved to the rear) that gave up and down motion. These could arguably be seen as parallels of the way a bird uses its tail, though there was less clear biomimetic inspiration in those cases.

Flying, then, was not a total failure for biomimetics, but the experience of the development of flying machines has an important lesson that uncritically following the natural approach is often not the best way forward. This was the case with the move away from flapping wings and would also occur with the refinement of the directly biomimetic wing-warping of the Wright brothers to become the more stylised, independent ailerons on the edges of a modern plane's wings. We can benefit from the inspiration provided by the world around us, but often it is necessary to adapt the approach taken to one that is better suited to our physical nature – or simply because we are able to find ways to go far beyond nature. The earliest flying machines may have been far less capable in the air than birds, but modern planes give us far greater speeds through the air than any natural example can provide.

There is one proviso to this limitation imposed by the differences between technology and nature though – what is not possible with one generation's technology does not necessarily remain impossible for ever. Such a leap typically requires both the availability of new ways of making something possible and a different viewpoint on the problem. It's not uncommon these days to mock the concept of 'thinking outside the box' as pointless management speak – yet it is very common for creativity to be limited by our assumptions of what is not possible.

Watching the 2021 film of the classic 1960s science fiction novel *Dune*, it is easy to doubt the realism of the insect-like ornithopters used to fly across the desert planet (though they are accurate to the description in the original novel). It's not just a matter of being certain that we will never see an ornithopter powered by a human turning a crank as suggested by Roger Bacon – any large-scale development of flying machines making use of flapping wings remains energetically impractical and has no benefit over conventional means of propulsion. However, if we could miniaturise the flying device, reducing its weight, the ornithopter approach may become practical again.

A company called Animal Dynamics has among its proof-of-concept designs what it describes as 'Skeeter' a 'flapping-wing micro-drone'. This is a tiny device weighing around 50 grams – about the weight of a slice of bread – that is based on the dragonfly. To some extent this can be seen as showing off what's possible, not necessarily providing a model for a commercial device. We know that drones with horizontally oriented propellors are now very stable and controllable – but it is possible that for very small applications, insect-like wings offer better versatility of flight control. The jury is out.

Seeing things differently

Another important illustration of the limitation of following nature too slavishly is in the way that four wheels are in many circumstances far more effective than animal-like feet as a way of enabling movement. A whole range of human innovations depend on rotary motion, yet this form of movement is rare in nature. It does exist in the biological 'motors' that power the whip-like flagella that act as propellors for some bacteria, but this was certainly not the inspiration of an unknown human innovator's first use of rotary motion. Yet had we tried to make early carts with feet, they would not have got far.

Another danger that biomimetic engineers face is that sometimes there is not a single factor involved in the way that a natural solution to a problem occurs – if the designers don't realise this, then it's possible to fail, because the way that the biomimetic solution is applied gains one benefit at the cost of introducing a new problem that was not expected. An extremely common observation in science is 'it's more complicated than we first thought', and often natural systems mislead us by appearing to be simpler than they really are.

Designers at Mercedes had exactly this problem when they came up with a concept car known as the Bionic. It's arguable that even if the biomimetic approach had worked out, the car would never have sold well as, by most aesthetic standards, it was extremely ugly. Yet the intention was to learn from the unusual shape of the yellow boxfish, which has a dramatic effect on its manoeuvrability through the water. Although it appears anything but streamlined, the fish's shape was thought to result in a dramatically reduced drag coefficient – the degree to which air resistance wastes energy – which would mean that copying the shape should reduce fuel consumption and make the car easier to accelerate.

Boxfish look clumsy and, well, boxy (just as the car did), but they are great at darting around the reefs in which they live. Pretty much uniquely for a fish, the boxfish has a stiff external carapace – so is far more like a vehicle than other fishy species. It seemed, then, an ideal biomimetic inspiration for a new car design. Unfortunately, there was a misunderstanding over the influence that the boxfish body shape had on drag, despite assumptions based on the fish's manoeuvrability. Just as the shape suggests, the fish offers

The yellow boxfish (*Ostracion cubicus*).
Norbert Potensky, CC BY-SA 3.0 via Wikimedia Commons

considerable resistance to moving through water – and, what's more, the rigid box carapace induces instability when the fish swims around. Despite the performance of some overpowered supercars, vehicle designers generally don't want their cars to veer all over the road.

It turned out that most of the fish's excellent ability to change direction and dart around came not from its shape at all – which made it less manoeuvrable than it otherwise would have been – but from the unusual operation of its fins, which were very flexible and could coordinate in various combinations to produce balletic turns and twists. Underlying this warning is the need to remember how evolution makes its changes. As we have seen, it is all too common to speak about evolution as if it were the outcome of design, but that is an inaccurate picture. It is a matter of random trial and error. Evolution does not produce the best solution – it produces a workable solution that may well be sub-optimal.

A lot of the blame for this misunderstanding of evolution arises from the watchmaker analogy, put forward by the 18th-century clergyman, William Paley in his book *Natural Theology*. Paley argued:

> In crossing a heath, suppose I pitched my foot against a stone, and were asked how the stone came to be there; I might possibly answer, that, for anything I knew to the contrary, it had lain there forever: nor would it perhaps be very easy to show the absurdity of this answer. But suppose I had found a watch upon the ground, and it should be inquired how the watch happened to be in that place; I should hardly think of the answer I had before given, that, for anything I knew, the watch might have always been there. Yet why should not this answer serve for the watch as well as for the stone? Why is it not as admissible in the second case as in the first? For this reason, and for no other, viz. that, when we come to inspect the watch, we perceive (what we could not discover in the stone) that its several parts are framed and put together for a purpose, e.g. that they are so formed and adjusted as to produce motion, and that motion so regulated as to point out the hour of the day; that, if the several parts had been differently shaped from what they are, of a different size from what they are, or placed after any other manner, or in any other order, than that in which they are placed, either no motion at all would have been carried on in the machine, or none which would have answered the use that is now served by it.

Paley makes the analogy of the watchmaker to support the religious idea more common at the time of there being a creator who has individually designed every organism and every aspect of nature. Many present-day religious believers do not take this viewpoint and accept that evolution is a natural process that does not require divine intervention – but it is easy to move away from the hand of a creator and still be influenced by the feeling that there is design involved, which implies some kind of holistic overview that simply isn't present when evolution is at work.

As a result, it was easy for the Mercedes designers to wrongly assume that the shape of the boxfish was the reason behind its manoeuvrability and stability, where in reality it was a hindrance that just happened to be more than compensated for by its remarkably effective tail and fins. Anyone indulging in biomimetics must be aware that nature's solutions are not always ideal, and that making use of just one factor may result in an outcome that is the exact reverse of what was first intended.

Not what it says on the tin

Perhaps the most interesting modern aspect of biomimetics is a phenomenon that appears to be straightforwardly biomimetic and is described as if based on a natural phenomenon, but in reality bears no resemblance to the natural ability it is named after. This is the field of artificial intelligence. It is hard to go a week without some new announcement of the wonders of AI. And science fiction has for around a hundred years prepared us for interaction with intelligent machines and robots. We've already seen how a mechanism for providing artificial intelligence – the neural network – is biomimetic. Yet the very term 'artificial intelligence' suggests that AI mimics the natural phenomenon of intelligence – and in the biomimetic sense, this is not the case.

You can argue exactly which organisms have some degree of intelligence, and it is quite difficult to pin down which of their actions are the result of intelligence as opposed to, say, conditioning. When, for example, a bird builds a nest, this is not usually seen as a reflection of intelligence as it seems pre-programmed to do it. However, unless we are being ironic, we have to accept that human beings have intelligence that they use in performing a wide range of tasks. For many years now, the field of information technology known as 'artificial intelligence' has been developing and has had some significant successes in carefully defined areas.

Outside of a relatively small number of enthusiasts, though, it is generally agreed that artificial intelligence bears no resemblance to what we might call general intelligence. AI-based devices are not general purpose and don't have similar capabilities as those provided by human intelligence. But there are some applications where they appear to mimic and indeed better human intellect, and it is worth exploring the most widely talked about of these to see how much they are indeed biomimetic of human intelligence. These are search engines, smart speakers, game-playing software and self-driving cars.

Ask Google

When the World Wide Web started to take off, it quickly became obvious that users were going to need help to navigate it. I remember getting hold of a copy of Netscape Navigator, the first mass-market browser, in 1995. At the time, you either found out about popular websites, such as an Australian botanical gardens site, because they were mentioned in magazines and laboriously typed in the address, or were referred from one site to another, as many had links to relevant sites of interest.

Early on, the descriptive model of the web that was most frequently used was a library. This wasn't a great model, but it was what we were used to as the only large repository of lots of documents. If you go into a big library, there is no way to find a particular book or document without help. The British Library, for example, holds over 170 million items. It's not like looking through someone's bookshelves. You need help to find what you want. Libraries have traditionally coped with the problem of finding a specific item in two ways. Firstly, the most frequently accessed books are on open shelves that are organised by topic, usually using a coding system known as the Dewey Decimal Classification, and then within topic by author.

However, this approach has its limitations. A library can only hold so many of its contents on open shelves – the big ones like the British Library keep most of their stock behind the scenes – and the 'by topic' approach has significant limitations.* Popular science could be treated as different to academic scientific fields, while science can be general or split, for example, placing books I've written into topics such as physics, maths, cosmology, technology and so on ...

---

* To be fair, arranging by topic and author is more logical than the way that the London bookstore Foyles used to arrange their books, which was by publisher – no one ever knows who the publisher is.

or some might be considered biography or history of science. And then there's the matter of someone like me who writes both fiction and non-fiction, who might also have books under science fiction and crime – each topic mentioned above (and many thousands more) has its own Dewey Decimal number.

Because of this potential for confusion and impossibility of providing all titles on open shelves, the second and more effective approach is to provide a catalogue or index – a way to look up books in various ways. Traditionally these would have been physical card indexes, probably with one each arranged by author and by topic, but with the introduction of computer databases it become possible to cut the data in many different ways.

This was, in effect, how guides to what was available on the web were first organised – using curated indexes such as Yahoo. But content grew in a way that made it impossible for human digital librarians to keep up with it. At the time of writing, there are between 1 and 2 billion websites online, featuring over 50 billion pages. The answer to finding our way around was the search engine. We are so used to the dominance of Google that it is easy to forget that it was not the first search engine. The biggest player before Google took over was AltaVista. This took a conventional data processing approach of building an index, simply providing links to sites based on keywords in a search. But Google moved the goalposts (and banished all competitors to a poor second place) by introducing artificial intelligence.

Rather than rely on looking at its indexes alone, Google enhances its responses to a query by making use of other people's work. When you get the answer to a search request, often formatted as a question, Google is not mechanically searching for the words in your question. So, for example, when I typed in 'How many items in the British Library?' to get the information above (and that is how I got it), Google did not just search on those words to decide what would come out at the top of my circa 369 million search results. Instead, it made use of how the results of other similar searches were treated and of how the pages it comes up with have been linked to by other pages (plus how reliable those other pages appear to be) in an extremely complex algorithm that deals with a couple of hundred different factors to produce what Google hopes are the best results.

This approach of apparently using experience to become better informed about the answer is what gives the process the appearance of being biomimetic of intelligence, prompting that label of 'artificial intelligence'. If Google were a person that somehow had been able to absorb all the information on the web and retrieve it at will, then it would be that person's intelligence that allowed it to understand what my question meant, interpret what I was looking for and decide how to best provide the appropriate answer. Google's AI appears to do something similar to this. But to understand why this isn't biomimetics in any conventional sense, we need to take a little diversion into what AI really *is* doing.

Artificial unintelligence

The reality is that by labelling these algorithms as artificial intelligence, we are indulging in something of a con. As sociology professor Elena Esposito points out in her book *Artificial Communication*, the term 'intelligence' implies the ability to understand something. When I ask my hypothetical person who can absorb all the information on the web a question, they understand what my words mean and provide me with what they understand to be the appropriate answer. AI can't do this, because an algorithm has no understanding of the questions it is asked.

Enthusiasts for artificial intelligence will point to AI's superhuman abilities – for example, learning to play the board game Go to a level that makes it possible to beat the best. But the Go-playing system AlphaGo, now owned by Google's parent company Alphabet, doesn't understand the game at all. Instead, it makes moves initially at random or adds in moves from old games, then gradually, by trial and error, learns which steps are most likely to result in a reward. It doesn't know that it's playing a game. It doesn't know what a game is. It has no idea that the 'questions' it is asked about what to do next are the result of someone else making a move. It is simply responding to a set of data to generate a new set of data – there is communication in the sense that data is being exchanged between the algorithm and the user, but there is no understanding. No intelligence.

Similarly, Google's search algorithm understands neither our question nor what is on the web, nor the material that it presents back to us. Its approach does not involve intelligence. Esposito refers to this not as artificial intelligence but artificial communication, illustrating that there is no biomimetic equivalence to intelligence simply because the AI can do a job that we use intelligence to carry out. It is not mimicking the approach taken by nature, it is using something entirely different. AI algorithms certainly communicate with us in a way that imitates human communication, but they don't mimic what lies behind the act of communicating.

So, are AI algorithms biomimetic? I would suggest that even though mimicking communication has a biomimetic quality, they aren't, but rather we dress them up in a way that makes them appear to be biomimetic. The whole conceit of referring to this as 'artificial intelligence' conspires to make something that isn't biomimetics look as if it is. Let's take a look at some of those other apparently biomimetic applications of AI to get a better understanding of why this is the case.

Hey, Siri

Many of us now have regular conversations with smart speakers or smartphones, appearing to chat with Siri, Alexa or Google Assistant. I have just asked Siri to tell me a joke. (It told me: 'I was trying to work out why the cricket ball kept getting bigger. Then it hit me.') Or I can ask it to convert a weight from kilograms to pounds, or to play some music. I can ask Siri to add a new appointment to my diary. And much more.

In a far more obvious way than is the case when doing a search on Google, this appears to be biomimetic. After all, the Google search is only mimicking my hypothetical person who knows everything that's on the web, who does not exist any more than Pierre-Simon Laplace's demon did. The demon was an imaginary being dreamed up by the 19th-century French philosopher which, in modern terms, we could say knows all the properties of everything in the universe. It knows the exact position of every atom, how it is moving and so forth. Laplace suggested that in a clockwork universe, where everything acts mechanically, obeying

Newton's laws, if his demon knew all of this it could predict the future perfectly.

Leaving aside the fact that quantum theory's dependence on probability that superseded the predictable Newtonian universe meant that Laplace's demon would fail, we know that Google doesn't really 'know' everything that's on the web. But when I ask Siri to tell me a joke, this is an interaction I might reasonably have with another human being (though I would hope that their jokes were better). I can make a conversational remark and Siri answers back, apparently just like a person.

In reality, I know it's not a genuine conversation, though all the time the companies behind these AI assistants are improving the software to make them appear more lifelike. It wasn't long ago that it was quite a shock when I started a conversation with 'Hey Siri' and then didn't say anything else, only to be prompted with a lifelike 'Hmm?' But just as Google is not truly intelligent, so the algorithms behind AI assistants are also not imitating the way that nature enables another human to hold a conversation.

Think about the true example of biomimetics, such as Velcro, or the paint that uses the lotus leaf effect to be self-cleaning in the rain. These products could not be confused with their biological inspirations. No one tries to fasten a piece of clothing using a burr from a burdock plant. The Velcro doesn't give the impression of being a burr while using a totally different mechanism to achieve similar results. It uses a similar *mechanism* to the burr to do something totally different. If anything, an AI assistant is anti-biomimetic. It does the same job, but the mechanism behind it is totally different to anything that occurs in nature.

Driving my car

The same limitations apply to self-driving cars as to AI assistants (which is why I suspect that it will take far longer for self-driving cars to take over the roads than many enthusiasts suggest). These vehicles give the impression of acting like an intelligent human driver, but in reality, that is not what they are doing. They do not imitate what makes a driver intelligent, but something entirely different provides the underlying mechanism.

In part this becomes very clear with the way that it is possible to fool such autonomous vehicles. Like the failed promised of intelligence in AI, the autonomy of these cars is only apparent – they still rely on the way that their algorithms function, which is not a matter of intelligence. Anything but. If I put a small, carefully designed sticker on a stop sign, which looks to human intelligence a bit like a QR code, a human driver would realise that this had nothing to do with the function of the road sign. But AI experts who know what they are doing can produce such a sticker that will fool a self-driving car into thinking that the stop sign is instead a speed limit sign, so that the car would plough straight on and cause an accident.*

When the accidents that autonomous vehicles have already had are examined forensically, it becomes clear that they are almost always due to a *lack* of intelligence. Self-driving cars have failed to recognise pedestrians if they

---

* AI experts don't, of course, want to make cars crash. But this vulnerability has been demonstrated, and there is no doubt that autonomous vehicle designers will face the same kind of challenges from malicious individuals as software designers do in preventing viruses and other malware causing disruption to our computers.

have been pushing a bike or carrying an odd-shaped object so they were no longer person-shaped – a situation that an intelligent person would have no problem recognising. Similarly, these systems have been fooled by street art (interpretating, say, a red blob as a traffic light) and have been confused by an overturned white truck, which was assumed not to be an obstacle because it did not have recognisable defining aspects in its outline and so the car drove straight into it.

In his book *Common Sense, the Turing Test and the Quest for Real AI*, the AI researcher Hector Levesque points out that the kind of machine learning used to train self-driving cars is excellent at dealing with everyday circumstances, but fails when it has to cope with an exceptional circumstance: something that bears limited resemblance to its training. Because human intelligence is based on an *understanding* of what is happening, a human is much more likely to be able to cope with unexpected situations. This is, in part, why an autonomous vehicle that manages reasonably well in a Californian city with a regular grid of straight roads and standardised junction layouts would be incapable of dealing with a winding British country lane, where pheasants are likely to fly out in front of traffic. Similarly, it would not be able to navigate across Swindon's notorious magic roundabout – something that looks terrifying, but which most humans can work out how to cross.

Of course, human drivers are fallible and are likely to have many accidents that an autonomous vehicle would be able to avoid. A self-driving car won't crash because it's looking at where the filling of the sandwich it was eating went when some of it fell out, or because it was trying to find a radio station. It won't lose attention because it's thinking of something other than driving. And that's great. But equally, the car lacks the intelligence to infer possible outcomes that are outside its training.

To give a practical example, a dual carriageway road near where I live has a series of traffic lights at junctions that incorporate pedestrian crossing points. No one has ever stepped out from one of those locations in front of my car. But when I pass one and, for example, I see children standing at the crossing, I am especially vigilant as to what might happen next. This vigilance does put me at a higher risk than an autonomous vehicle of driving into the car in front of me (at relatively low speed), but it makes me more likely to anticipate an accident that's about to occur if one of those children does something child-like. I know, for example, that if a child is bouncing a ball at the side of the road, this very much increases the risk that they will run out – the car doesn't know this. And my potential error would be far less tragic than the car's.

Similarly, I observe the road in front of me through the lens of intelligence. I know what a traffic light signal is. I know what a truck looks like, whichever way up it is. I know what people look like from any direction and I know how they are likely behave. All this gives me an advantage in low probability but high-risk circumstances.

It is entirely possible that, as the vendors claim, self-driving cars would cause fewer accidents than human-driven cars (though I suspect we are a long way off this being the case on all roads and in all driving conditions). However, we should not mistake the apparent ability of these cars to be biomimetic of human drivers for a true representation of what they are doing. They are not – their ability to drive is based on something entirely different from intelligence.

What about robots?

In essence, a self-driving car *is* a robot. And like these vehicles, the vast majority of real-world robots are not anthropomorphic. However, in part thanks to science fiction, we have an expectation of robots to provide the ultimate in biomimetics, not only displaying artificial intelligence, but mimicking human or animal physical form and the ability to move like a living thing too.

All the arguments above about using something different from intelligence to make decisions also applies to these robots. In their physical abilities, some attempts have been made to develop artificial muscles that work a little like biological muscles and so would be biomimetic – but in practice these tend to be concept demonstrators. In practice, it is almost always better to provide a function that a biological organism performs using a different mechanism. Just as machine learning does not equate to intelligence, so, for example, the ability of a robot such as Honda's ASIMO to walk on two legs does not use the same internal mechanism as a human.

It's possible to argue that there is a higher level of biomimetics going on here. Non-biomimetic wheels, for example, are far better than legs at moving down a highway at 70 miles an hour. But legs can offer robots advantages over wheels on some kinds of terrain. So, if we take the design of robots vaguely based on humans or other animals at as high a level as 'using legs', then there is the potential for a degree of biomimetics – but nothing like the biomimetic weight that is implied by the humanoid appearance of some robots.

Although we certainly can benefit by copying some aspects of nature, we usually do better when we use nature as a jumping-off point, from which we can employ human ingenuity to develop the idea and put it into action, often exceeding what is possible through unmodified mimicry of biology or other aspects of the natural world. With this in mind, we can go on to look at possibilities for the future of biomimetics.

FACTS AND FUTURES 8

It may seem from the previous chapter that the future of biomimetics is limited. But although we do need to always be careful not to oversell the potential of any particular biomimetic approach, we have only scratched the surface of the potential of biomimetics. The field is developing all the time. And the purpose of this book is not to detail every possible biomimetic product or concept, but rather to highlight the kinds of application that typify different kinds of biomimetic inspiration.

A different view

This book has turned out to be very different from the one I originally intended to write. The reality of what has happened in this field is at odds with the glossy marketing that refers to it (or, for that matter, practically every book that has been written about it). This lack of following expectation is not a bad thing. Talk to any scientist, and they will tell you that science gets interesting when the world does not behave the way we expect it to. Based on this metric, biomimetics and its implications for our understanding of the use of naturally inspired science and technology in new products and design is very interesting indeed.

The inevitable exemplar of biomimetics is Velcro, which is why it appeared in the opening chapter. Everyone knows what Velcro is. Most people will have some hook and loop fasteners in their home. Yet, as we have explored the biomimetic world, it has become clear that Velcro is anything but typical of biomimetic products. It's a bit like using Usain Bolt as a representative example of a weekend jogger.

Most biomimetic products are either successful in a niche – good examples being lotus effect paint and its mirror image self-cleaning glass – or make good stories but haven't really had any impact at all. A major factor is cost. Many biomimetic inspirations may be superior to the equivalent conventional product, but it is hard to justify the additional cost of the more sophisticated design for the benefits gained. Arguably, Velcro is relatively unusual because it is cheap and offers a kind of fastening that is different to, and often better than, alternatives. Where lotus effect paint is indeed a better house paint, and self-cleaning glass is a better kind of glass to the regular variant, these are improvements that don't affect the prime use of the product. By contrast, Velcro is totally different to buttons or zips and has many applications that would be impossible with other types of fastener.

Perhaps the true exemplar of biomimetics is the yellow boxfish-based Mercedes car, Bionic. Not because it's a good example of the benefits of biomimetics – it really was hopeless. But rather it is significant because Bionic was a concept car. Anyone familiar with the automobile industry will know how often concept cars bear little or no resemblance to the company's production models. Many of the ideas they incorporate prove to be totally useless and are rapidly dropped. Others do make it to market but are so heavily modified that they are almost unrecognisable. But concept cars can still act as inspiration for the future.

It's not that biomimetics is doomed. It will continue to inspire many new concepts. Some will become niche products. Others may genuinely have a chance to become the next Velcro. However, we have to understand that it's rarely about seeing something in nature and making direct use of an equivalent engineered product to produce the same effect (as was the case with Velcro). Instead, it's more often about producing starting points – jumping off points to do something different.

What's next?

There have been plenty of books written that make predictions about future technology and one thing they tend to have in common is that they get things wrong, often dramatically so. Looking back on what was probably the biggest futurology title of all time, *Future Shock*, Alvin Toffler's 1970 bestseller, there's plenty that went wrong with his predictions of the 21st century. To see why, it's useful to take a look at one specific area of prediction in that book – disposability.

Reflecting the change, particularly in America, that had brought in more and more of a throw-away society by the late 1960s, Toffler envisaged a future where this approach was taken to the extreme. Apparently, in 1970, paper dresses were all the rage (I can't say I remember this), and wear-once-then-throw-away clothes were something Toffler assumed would become the norm. Realistically, paper clothes were always a non-starter as anything more than a gimmick – certainly in Manchester, say. But it is true that the current young generation does think of clothes as more short-term purchases compared with a generation that bought clothes and kept them until they wore out. (My raincoat is over 30 years old and still going strong.)

However, what Toffler missed is the way that an awareness of green issues, and the enabling power of the internet, would become a natural background to life. While the younger generation might not hang on to clothes the way some older folk do, they also don't just throw them away. Instead, they resort to recycling them, donating them to charity shops or selling them on online services such as eBay and Depop. This is in part driven by increasing awareness of the environmental impact of cheap clothing (though there is certainly a financial incentive too). And the same goes for many of our everyday things. Yes, we do change some products a lot more than we used to, but equally we tend to recycle them, ideally for money. It would have seemed crazy in 1970 to change your phone every two years, say (it would, of course, have been a landline phone), but when we do make the change, we trade in the old one or sell it.

What this illustrates is how difficult it is to spot how social and technical changes will be reflected in the way we live in the future – and the same applies to biomimetics. As yet, there have been few true breakthrough biomimetic products, but that doesn't mean that something significant will not emerge.

Learning from life

What seems likely is that as we get a deeper understanding of the mechanisms that make biological organisms work, we will have more potential to replicate in our technology some of the very special abilities of living organisms. We have seen, for instance, some early moves to apply the self-healing ability of biological organisms to concrete – but the potential would be even greater, say, with self-healing plastics or glass equivalents. Could this mean no more cracked phone screens?

Two other key features of living organisms are the abilities to harvest energy from the environment and to reproduce. Although we are seeing increasing use of photovoltaic cells to produce energy from sunlight, there are wider possibilities open to future, more biomimetic, technology perhaps making use of chemical energy to produce power from waste or biological material. Grass cuttings in the future, for example, might not just produce compost but energy too.

Reproduction might seem a step too far, but there have been suggestions in the past of the possibility of machinery that can produce copies of itself, given appropriate energy and raw materials. This is obviously attractive in some circumstances. A classic potential application is the von Neumann probe, which the Hungarian–American polymath John von Neumann suggested was the future of interstellar exploration.

The idea here is that because of the vast distances between stars, by far the most efficient way to explore interstellar distances is not to send out Captain Kirk in the USS *Enterprise*, but rather to build probes that would travel to another planet, where they would replicate themselves and send out more probes to other planets, becoming distributed across a wide stretch of the galaxy over time.

A more down-to-earth reflection of nature is coming in mechanisms that remove carbon dioxide directly from the air. Although not using technology that is similar to the carbon capture of plants, such 'direct air capture' systems can be seen as inspired by plants in the way that they pull carbon dioxide out of the atmosphere and store it away. This is a lot harder to achieve than carbon capture and storage systems that prevent carbon dioxide being added to the atmosphere from power plants and industry, but it can make a significant difference as it is carbon negative, rather than carbon neutral. A number of approaches using specialist filters or chemical reagents have been tried, but all lock onto carbon dioxide and enable it to be stored (or used) away from the atmosphere. At the time of writing, there are only a handful of such direct air capture plans, but many more can be expected.

Viruses and worms

In a way, there is a similarity between von Neumann probes and computer worms. Computer viruses and worms are at the very least nominally biomimetic – and worms in particular, which are designed to spread from computer to computer automatically, reproduce in a biomimetic fashion. If we can be certain of one thing in the future, it is that computer (and phone) viruses and worms will not go away. It's worth reflecting on the very first computer worm and its remarkable impact to consider what may come in the future.

This antique of the malicious computing world infected the Arpanet, the predecessor of the internet, which by 1988 had around 60,000 computers linked to it.

In late 1988, operators running some of these computers noted that their machines were slowing down. It was as if many people were using them, even though, in practice, their loads were light. Before long, some of the computers were so slow that they had become unusable. And like a disease, the problem spread from computer to computer.

To begin with, operators tried taking individual computers off the network and restarting them – but soon after reconnecting, the clean machines started to slow down again. In the end, the whole Arpanet had to be shut down to flush out the system. Imagine the equivalent happening with the current internet. Shutting down the whole thing, including everything connected to it. The impact on global commerce, education and administration would be colossal. Thankfully, the Arpanet of the time was relatively small and limited to academia – but its withdrawal still had a serious cost attached.

The apparent attack on the Arpanet was the result of an innocent error made by a graduate student at Cornell University by the name of Robert Morris. Because the Arpanet was growing organically it was hard to judge just how big it was. Morris had the idea of writing a program that would pass itself from computer to computer, enabling a count of hosts to be made. In essence, he wanted to undertake a census of the Arpanet – a perfectly respectable aim. But the way he went about it would prove disastrous.

Morris had noted a number of issues with the way the UNIX computers that dominated the Arpanet worked. The sendmail program, used by UNIX to transfer electronic mail from computer to computer, allowed relatively open access to the computers it was run on – while at the same time, in the free and easy world of university computing of the period, many of those who ran the computers and had high-level access had left their passwords blank. It proved easy for Morris to install a new program on someone else's computer and run it. His worm was supposed to spread from computer to computer, feeding back a count to Morris.

There's no doubt that Morris knew he was doing something wrong. When he had written his self-replicating worm, instead of setting it free on the Arpanet from Cornell he logged on to a computer at MIT and set the worm going from there. But all the evidence is that Morris never intended to cause a problem. It was because of a significant error in his coding that he would be landed with a criminal record.

When the worm gained access to a computer, its first action would be to check whether the worm program was already running there. If it was, there was no job to do and the new copy of the worm shut itself down. But Morris realised that canny computer operators who spotted his worm in action would set up fake versions of the program, so that when his worm asked if it was already running, it would get a 'yes' and would not bother to install itself. Its rampant spread would be stopped, and its survey mission would fail.

To help overcome this obstacle, Morris added a random trigger to the code. In around one in seven cases, if the worm got a 'yes' when it asked if it was already running, it would install itself on the computer anyway and would set a new copy of itself in action. Morris thought that this one-in-seven rule would keep the spread of his worm under control. He was wrong.

The internet, and the Arpanet before it, is a particular kind of network, one that often occurs in nature. Because it has a fair number of 'hubs' that connect to very many other computers, it usually only takes a few steps to get from one location on the network to another. What's more, it was designed from the beginning with redundancy. There was always more than one route from A to B, and if the easiest route became inaccessible, the system software would re-route the message and still get it through. Bearing in mind its military origins, the builders of the Arpanet believed that at some point in the future it was likely that someone would attempt to take out one or more parts of the network. The network software and hardware made this less of a problem.

The feedback of virus attacks

The combination of the strong interconnectedness of the network and the extra routes to withstand attacks meant that Morris' one-in-seven rule allowed a positive feedback loop to develop. This is a highly biomimetic phenomenon. Natural droughts, for example, involve a positive feedback loop. If rainfall drops off and plants die off, there are fewer plants to put water vapour into the atmosphere – which means the drought gets worse. And so on. On a more positive note, fruit ripening is often also driven by positive feedback. For example, as the first fruits ripen on an apple tree, the ripe fruit give off ethylene gas, which encourages other apples to ripen, so the fruit tend to ripen together more effectively.

The same kind of effect was happening with the computers on the Arpanet as Morris' worm took hold. The more the worm was installed on a computer, the more it tried to pass itself on to other computers – and the more copies of the worm that were installed on the computer. Before long, hundreds and then thousands of computers were running more and more copies of the worm. And each copy that ran slowed the computer down until the machines ground to a halt.

Just how dramatic this effect was can be judged by a time-based report on the status of the first computer to be infected, a DEC VAX minicomputer at the University of Utah. It was infected at 8.49pm on 2 November and twenty minutes later began to attack other computers. By 9.21pm, the load on the VAX was five times higher than it would normally be – there were already many copies of the worm running. Less than an hour later, there were so many copies running that no other program could be started on the computer.

The operator manually killed all the programs. It then only took twenty minutes for the computer to be reinfected and to have ground to a halt again. The only way to avoid repeated reinfection was to disconnect it from the network. Ironically, Clifford Stoll, one of the operators responsible for a computer taken over by Morris' worm, rang up the National Security Agency and spoke to Robert Morris' father, who worked there. Apparently, the older Morris had been aware of the flaw in sendmail that the worm used for a number of years. But at the time of the call, no one knew that it was Morris' son who had started the worm on its hungry path through the net.

The Arpanet survived the unintentional attack and Morris (junior) became the first person to be convicted under the Computer Fraud and Abuse Act, receiving a hefty fine, community service and probation. Morris' worm was, by the standards of modern computer viruses and worms, very simple. It wasn't even intended to cause a problem. But it shows us how a biomimetic approach to a simple problem can have a profound and unexpected impact.

More to come

Other biomimetic applications we may see more of are opportunities for developing the swarm concept and making use of a bat-like echo locator to assist the visually impaired, or those who need to operate in darkness. Despite being a very expensive way to move on the flat, robotics that make use of animal locomotion, such as Boston Dynamics' intriguing robotic dogs or the École Polytechnique Fédérale de Lausanne's robots based on the walking motion of salamanders, could give us robots better able to take on varied terrain.

Similarly, we can see biomimetic possibilities from taking a very specific aspect of a natural phenomenon and applying it to rather different uses – just as the kingfisher's dive inspired the Shinkansen train's nose. For example, researchers at the University of Manchester have been looking at why birds don't fall over when they attempt to take flight. When a bird launches itself, it is often from a standing start, and this isn't easy.

This is because the bird has to lean forward to get their first flap in, and this could easily result in a nose-dive into the ground. To avoid this, birds perform a manoeuvre referred to as 'pitching upwards', a kind of jump and flex motion that is an integral part of their take-off process. The hope is that understanding this will give guidance on how to design robots that can jump without falling over, making it easier to cover ground that is covered in debris, such as in the aftermath of an earthquake or bombing.

But these are stabs in the dark. Perhaps the biggest lesson from both Velcro and the Arpanet worm are that the really successful biomimetic products tend to be simple and not expensive to implement. So many biomimetic projects to date have involved such complexity, or extra expense that is difficult to justify for the benefits gained.

The result of exploring biomimetic design will often have limited applicability, in part because it can be extremely costly to reproduce what nature manages so effortlessly. But there will be new wonders – and we can continue to marvel at the ways in which the vast scale of nature's randomised test-and-fail system can produce solutions to everyday problems.

Biomimetics is not going away any time soon.

FURTHER READING

Books

Brian Clegg, *Everyday Chaos: The Mathematics of Unpredictability, from the Weather to the Stock Market* (Cambridge: MIT Press, 2021) – more detail on fractals and Sierpiński gaskets.

Brian Clegg, *The First Scientist: A Life of Roger Bacon* (London: Constable, 2013) – exploration of Roger Bacon's life and work.

Elena Esposito, *Artificial Communication: How Algorithms Produce Social Intelligence* (Cambridge: MIT Press, 2022) – explores why the term 'artificial intelligence' is so misleading. Heavy going, but informative.

Peter Forbes, *Dazzled and Deceived: Mimicry and Camouflage* (New Haven: Yale University Press, 2009) – Forbes introduces natural examples of mimicry and camouflage and how humans have learned from them.

Peter Forbes, *The Gecko's Foot: How Scientists are Taking a Leaf from Nature's Book* (London: Harper Perennial, 2006) – Forbes gives a detailed, lyrical description of a handful of major biomimetic applications.

Diarmuid Jeffreys, *Aspirin: The Remarkable Story of a Wonder Drug* (London: Bloomsbury, 2005) – detailed history of aspirin and its uses.

Hector J. Levesque, *Common Sense, the Turing Test and the Quest for Real AI* – (Cambridge: MIT Press, 2017) – demonstrates the limits of machine learning that arise because its mechanisms are different from human reasoning and intelligence.

Clifford Stoll, *The Cuckoo's Egg* (London: Pan Books, 1991) – the story of the biomimetic Arpanet worm.

Alvin Toffler, *Future Shock* (S.I.: Random House, 1970) – an attempt to predict 30 plus years into the future, showing the dangers of predicting technology.\

Laurie Winkless, *Sticky: The Secret Science of Surfaces* (London: Bloomsbury Sigma, 2022) – contains a detailed exploration of the gecko's foot, including several recent developments.

Online

Animal Dynamics *flapping-wing micro-drones* – https://www.animal-dynamics.com

Bcomp powerRibs *bodywork material for motorsport* – https://www.bcomp.ch/solutions/motorsports-bodywork/

Boston Dynamics *robotic dogs and more* – https://www.bostondynamics.com

Continental *frog-inspired tyres*: https://www.continental-tyres.co.uk/car/tyres/contiwintercontact-ts-850

École Polytechnique Dédérale de Lausanne *Salamandra robotica* – https://www.epfl.ch/labs/biorob/research/amphibious/salamandra/

Eden Project – https://www.edenproject.com

Fractal sunshades – https://en.losfee.net/environmental-product-design

Frog-inspired tyre treads – https://www.uakron.edu/im/news/with-the-help-of-frogs-graduate-student-wins-award-for-tire-tread-research

Gecko Adhesion Research Group, University of Akron –
https://uagecko.wordpress.com

Gecko tape – *Geim's early version* – https://physicsworld.com/a/
gecko-tape-sticks-with-polymer-fibres/

Jumping robots – *University of Manchester study of bird launching* –
https://sites.manchester.ac.uk/biomimetics/research/

Lindisfarne lookout tower – http://www.icosis.co.uk/
the-lookout-tower-lindisfarne/

Lotusan *self-cleaning paint* – https://www.stocorp.com/
sto_products/stocolor-lotusan-low-voc/

Magic Roundabout – *Swindon's unique five-roundabout junction* –
https://en.wikipedia.org/wiki/Magic_Roundabout_(Swindon)

Materials for Life – *self healing concrete test in South Wales* –
https://www.frontiersin.org/articles/10.3389/
fmats.2018.00051/full

Mercedes Benz Bionic car – https://www.mercedes-benz-media.
co.uk/en-gb/releases/223

Morphotex *fabric with photonic colouring* – https://www.
nanotechproject.tech/cpi/products/morphotex-r-fiber/

ORNILUX *bird protection glass* – http://www.ornilux.com

Oxford Biomaterials – *spider silk-inspired medical technology* –
http://www.oxfordbiomaterials.com

Photonic cosmetics – *L'Oréal's exploration of photonic crystals* –
https://www.cosmeticsdesign.com/Article/2005/08/04/L-Oreal-
focuses-on-photonics-to-enhance-colour-cosmetics

Pilkington Activ *self-cleaning glass* – https://www.pilkington.com/
en-gb/uk/householders/types-of-glass/self-cleaning-glass

Spintex *spider silk-inspired artificial silk* – https://www.spintex.co.uk

Velcro *Don't say Velcro campaign* – https://www.velcro.co.uk/
dont-say-velcro/

Velcro *patent* – https://patents.google.com/patent/US3009235

Whalepower Corporation *humpback whale-inspired turbine blades* –
https://whalepowercorp.wordpress.com

INDEX

We inhabit a planet in peril. Our once temperate world is locked on course to become a hothouse entirely of our own making.

Hothouse Earth provides a post-COP26 perspective on the climate emergency, acknowledging that it is now impossible to keep this side of the 1.5°C climate change guardrail. The upshot is that we can no longer dodge the arrival of disastrous – and all-pervasive – climate breakdown that will come as a hammer blow to global society and economy.

Bill McGuire explains the science behind the climate crisis and presents a blunt but authentic picture of the world bequeathed to our children and grandchildren; a world already glimpsed in today's blistering heatwaves, calamitous wildfires and ruinous floods and droughts. This picture is one we must all face up to, if only to spur genuine action to stop a harrowing future becoming a truly cataclysmic one.

ISBN 978-178578-920-5 (paperback)
ISBN 978-178578-921-2 (eBook)

Music is shaped by the science of sound.

How can music – an artform – have anything to do with science? Yet there are myriad ways in which the two are intertwined, from the design of instruments and hi-fi systems to how the brain processes music.

Science writer Andrew May traces the surprising connections between science and music, from the theory of sound waves to the way musicians use mathematical algorithms to create music.

The most obvious impact of science on music can be seen in the way technology has revolutionised how we create, record and listen to music. Technology has also provided new insights into the effects that certain music has on the brain, to the extent that algorithms can now predict our reactions with uncanny accuracy, which raises a worrying question: how long will it be before AI can create music on a par with humans?

ISBN 978-178578-991-5 (paperback)
ISBN 978-178578-990-8 (eBook)